Basic
Engineering
Craft Studies

Contributors

G. S. Birkby
P. H. M. Bourbousson
H. Carter
V. Duckworth
K. C. Jackson
F. Mundy
E. S. Scragg
A. V. Smith
E. W. Webb

Illustrators

L. D. Garrett
C. Holmes
Mary Stevens

Basic Engineering Craft Studies

Advisory Editors

P. H. M. Bourbousson
> Stevenage College of Further Education
> Stevenage

R. Ashworth
> Stannington College of Further Education
> Sheffield

Mechanical (03)
> Complementary book to be used in
> conjunction with Base book—
> General (01)

Butterworths

First published in 1971
    Reprinted 1977, 1979, 1981

© Butterworth & Co (Publishers) Ltd, 1971

ISBN 0 408 00057 0

Set, printed and bound in Great Britain by
Fakenham Press Limited,
Fakenham, Norfolk

Preface

This book, combined with the companion Base book (General 01), has been written by a team of technical college lecturers, each of whom is a specialist in a particular field of craft course teaching, for students studying for the City and Guilds of London Institute 200 Course in Basic Engineering Craft Studies (Part I) – Mechanical Bias (01/03).

The Base book covers the 01 material and this volume completes the material as required by the 03 part of the syllabus.

The complete syllabus has been divided into six sections, i.e. A Calculations, B Material Removal, C Materials, D Measuring and Marking Out, E Fabrication and Welding, F Electrical, each section being subdivided and allocated a reference number. This topic reference number is used both in this Complementary volume and in the Base book, e.g. Section E7 Gas Welding is introduced in the Base book and completed under the same reference (E7) in this volume. Some topics are not introduced in the Base book, e.g. E11 Forging and Forming, and in these cases the complete material appears in this volume.

In the writing of these books particular emphasis has been placed on presenting as much information as possible diagrammatically. The diagrams should provide many discussion points between student and lecturer and will assist the student in his understanding of Engineering.

At the end of each section, with the exception of Section A, is a series of questions which the Editors feel will be useful to the student and assist him to comprehend the material in the books.

Apart from students studying for the above-mentioned course, these books will prove useful for adults attending government training centres, and for men employed in industry who wish to extend their knowledge of basic engineering.

The welding symbols on pages 66 and 67 have been extracted from B.S. 499 (Part 2): 1965 and the colour codes for the contents of pipes, on page 123, from B.S. 1710:1960, by permission of the British Standards Institution, 2 Park St., London W.1.

P.H.M.B.
R.A.

Contents

A4   Angles

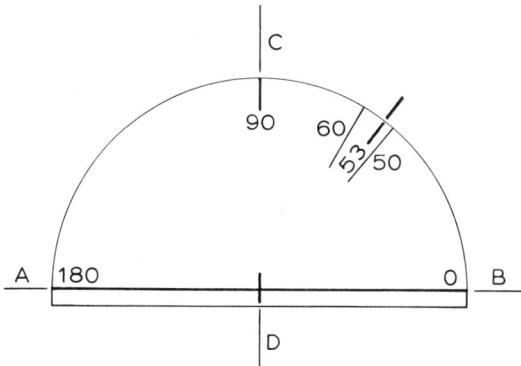

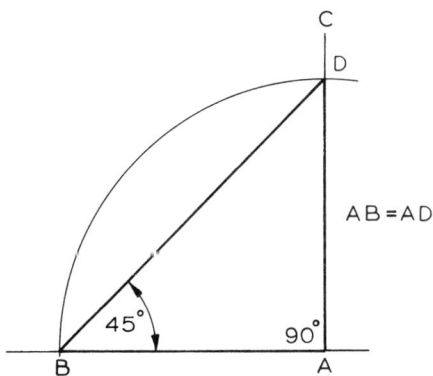

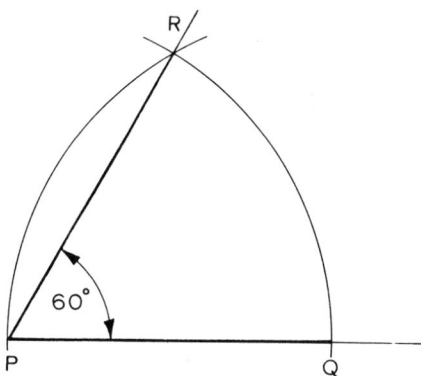

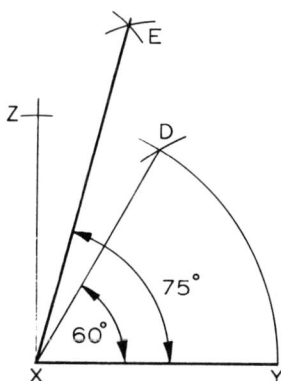

Setting Out Angles With and Without a Protractor

When setting a protractor in position for marking off an angle, care must be taken to line up the base line, i.e. the line marked 180–0, exactly on top of the line on the work from which the angle is being marked. As a check, where possible line up the 90° line on the protractor with a line at right-angles to the base line on the work—AB and DC in the illustration.

When marking off, take care to view the gradation on the protractor from directly above to avoid the effect of parallax.

Standard 45° and 30° squares may be so arranged as to give angles in multiples of 15°. Some of these are shown in Section A4, page 19 of Base book.

The second illustration shows the construction of a 45° angle. Starting with AD, use A as centre to describe a 90° arc, the radius being AB. AD and AB are at right-angles. The line BD is at 45° to both AB and AD.

The construction of an angle of 60° is given opposite. Starting with PQ, use P as centre to describe an arc having a radius equal to PQ. Keeping exactly the same radius, and with Q as centre, make an arc from P. Let the arcs intersect at R; then PR is at 60° to PQ.

To construct a line at 75° to another the construction in the bottom illustration may be used. XZ and XY are at right-angles. First, construct XD at 60° to XY; then, bisect angle DXZ. Then, angle EXY = 75°.

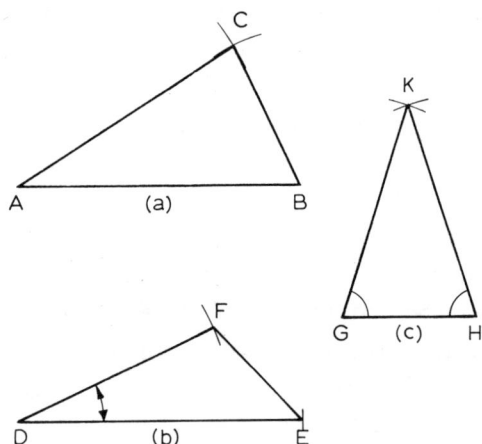

A5   Plane Figures   A5

Triangles

Scalene—(a) and (b) in top illustration.

(a)   With the lengths of the sides given:
Draw AB.
Centre A, draw an arc, radius = AC.
Centre B, draw an arc, radius = BC.
Join A to C, and join B to C.

(b)   With two sides and included angle given:
Draw DE.
Mark off angle EDF.
Measure DF to size.
Join E to F.

Isosceles—(c) in top illustration.

Draw GH to length.
Set dividers or compasses to length of side
GK or HK. Mark off arcs from centres G
and H, thus forming point K.
Join G and H to K.

Right-angled—second illustration.

Set off the sides as in a scalene triangle. The
proportions given here produce a right-angle
as shown. There are other sets of sizes giving
a right-angle, e.g. 5, 12, 13.

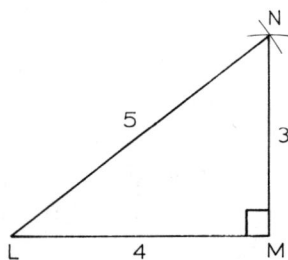

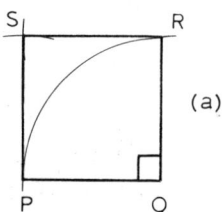

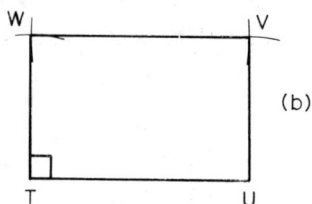

Squares and Rectangles

(a)   The square starts with PQ being drawn
and marked off to length. QR is drawn
at right-angles as a construction line.
Set compasses to PQ and make an arc
up to form R. With the same radius use
P and R as centres to make arcs for
point S. Complete the square.

(b)   To start the rectangle draw two lines at
right-angles from which TU and TW
may be marked off. Then using U as
centre, radius TW, mark an arc in the
region of V. With centre W, radius
TU, mark an arc to give the point V.
Join up to give the rectangle TUVW.

Parallelogram

Start with the angle XOZ, then proceed as in
the rectangle.

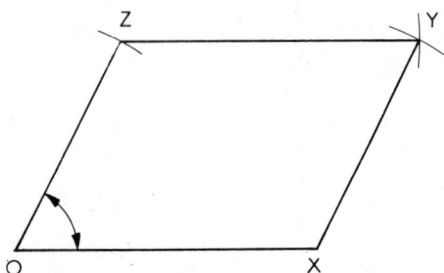

Hexagons

On this page various methods of constructing hexagons are shown. Although in any diagram the 'corners' may be shown at top and bottom, they could be on a horizontal line. The student will find it useful to try to do this.

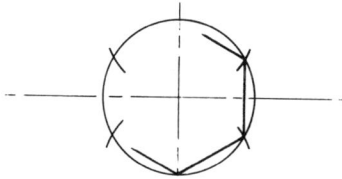

Draw a circle. Keep the compasses set at exactly the same radius and step it out round the circle—it will go six times. Join the points to form the hexagon.

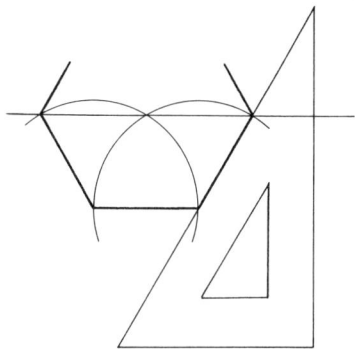

Generally, it is most convenient to start with the horizontal base, but the construction is similar if begun with a vertical line. Draw the base line to size. Draw a line at 60° as shown by the square. Measure off this line to size, or swing round an arc as shown. Continue with square and marking off until the hexagon is complete.

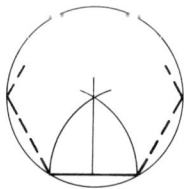

In this case the length of the side is given. Draw the 'base' to size. Swing arcs up from the ends of the base. The point of intersection is the centre of a circle whose radius is equal to the length of a side. Draw round the inside of the circle using the 60° square until the hexagon is complete.

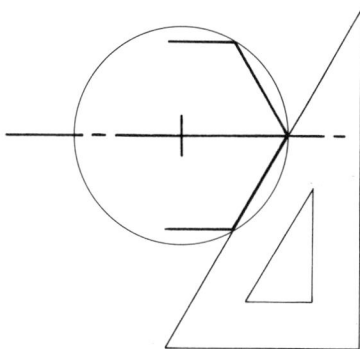

The object here is to draw a hexagon inside a circle.
Draw a horizontal diameter across the circle. From one end of this diameter use the 60° square to make the first side of the hexagon. By turning the square over, another side may be drawn from the other end of the diameter. Similarly, all the sides may follow, the top and bottom being drawn horizontally.

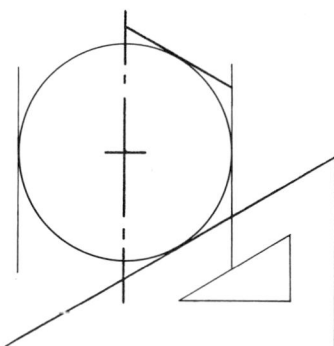

The hexagon is drawn here on the *outside* of a circle. The dimension given is *A/F*, i.e. *Across Flats*. After drawing the circle simply draw lines, lightly at first, at the appropriate angles to make the hexagon.

3

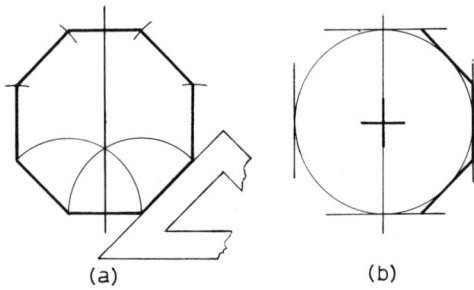

(a)   (b)

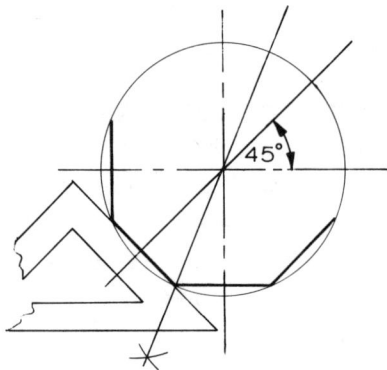

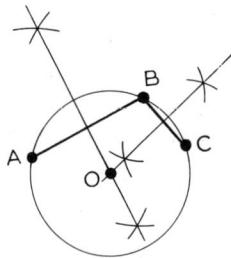

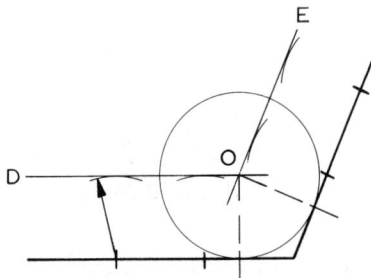

Octagons (eight-sided figures)

The constructions (a) and (b) are similar to those used for hexagons. In (a), draw one side to size, draw a line at 45°, mark off to size; continue with a vertical line. Proceed until all the sides are drawn.

In (b), with the A/F dimension given, draw the circle of that diameter and with a 45° square complete the polygon.

To draw an octagon within a circle, i.e. having the dimension 'over the corners', start with a diameter (say vertically). Draw a diameter at 45° to the first. Then bisect the 45° angle. This line will cross the circle at a corner of the octagon. Proceed to draw successive lines, using the 45° square, to complete the polygon.

Circles

To construct a circle to pass through three points, the lines joining the points may be considered as chords of the circle.
Construct the perpendicular bisectors of AB and BC. (Equal arcs from each end above and below each line.) The point of intersection of the bisectors is the centre of the circle, O. The radius is OA or OB or OC.

The circle shown left is drawn so as to touch the straight lines.
The lines DO and EO are parallel to the given lines. This is done by making small arcs, with centres on the given lines and radius exactly that of the circle. Draw through the arcs. The intersection gives the centre, O.

Exercises

Draw the triangles where:
1. AB = 70, AC = 100, BC = 110 mm.
2. AB = 65 mm, AC = 80 mm, angle BAC = 40°.
3. AB = 100 mm, angle BAC = 35°, ABC = 72°.
4. The sides are 125, 100 and 75 mm.
5. AB = 70 mm, AC = BC = 90 mm.

Draw the hexagons where:
6. The sides are 40 mm.

7. Across corners = 100 mm.
8. A/F = 120 mm.
Draw the octagon:
9. With sides of 40 mm.
10. Where A/F = 100 mm.
11. Within an 80 mm diameter circle.

12. Construct a circle to pass through points A, B and C, where AB = 70, BC = 40 and AC = 90 mm.
13. Construct an 80 mm diameter circle to touch two lines meeting at 130°.

A8   Three-Dimensional Figures   A8

Volume and Surface Area of Rectangular Prisms, Cubes and Cylinders

Rectangular Prisms

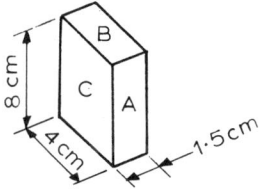

Surface Area

There are six surfaces, two like A, two like B and two like C.

Area A $= 8 \times 1.5$

So, the two As have an area of $2 \times 8 \times 1.5$, i.e. $16 \times 1.5 = 24$ cm².

Areas B $= 2 \times 4 \times 1.5$
$= 12$ cm²

Areas C $= 2 \times 8 \times 4$
$= 64$ cm²

Total surface area $= 24 + 12 + 64$
$= \underline{100 \text{ cm}^2}$.

Volume

The volume of any prism may be calculated by multiplying the area of cross-section by the length or height. In the illustration above, the face B may be taken as the cross-section.

Then, volume $=$ area of B $\times 8$
$= 4 \times 1.5 \times 8$
$= \underline{48 \text{ cm}^2}$.

Cubes

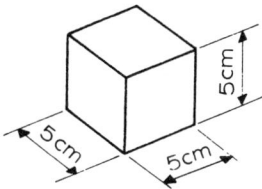

Surface Area

There are six equal faces; the area of any face is the square of the length of a side. Thus, in the illustration, the total surface area

$= 6 \times 5^2$
$= 6 \times 25 = \underline{150 \text{ cm}^2}$.

Volume

The volume is the cube of the length of a side. Thus, in the cube shown,

Volume $= 5^3$
$= 5 \times 5 \times 5 = \underline{125 \text{ cm}^3}$.

Problems

In Problems 1 to 7, find the surface area.

1   A rectangular block of copper, 24 mm long, 15 mm wide and 5 mm thick.
(1 500; 1 110; 1 050; 910)

2   A steel plate, 1·4 mm thick, 15 mm wide and 20 mm long.
(698; 728; 648; 778)

3   A tank (open top), 3·5 m long, 1·6 m wide and 0·75 m high.
(10·5; 11·25; 12·5; 13·25)

4   A tray, 40 cm wide and 55 cm long. It has vertical sides 8 cm high.
(3 720; 3 800; 4 000; 4 220)

5   A shield for an assembly is open at both ends. It is 45 mm long and the dimensions of the end are 30 mm and 12 mm.
(4 280; 3 920; 3 780; 3 250)

6   A cube with sides of 8 cm.
(254; 294; 338; 384)

7   A cube with sides of 2·4 m.
(34·56; 30·6; 28·04; 24·32)

In Problems 8 to 10, find the volume.

8   A 1 m length of aluminium, 3 cm wide and 0·5 cm thick. Answer in cm³.
(100; 120; 150; 180)

9   A cube with sides of 15 cm.
(3 125; 3 375; 3 525; 3 750)

10 A cubicle, 6·5 m long, 4 m wide and 3·5 m high. Answer in m³.
(95; 91; 83; 80)

Answers: **1** *b*; **2** *a*; **3** *d*; **4** *a*; **5** *c*;
6 *d*; **7** *a*; **8** *c*; **9** *b*; **10** *b*.

Cylinders

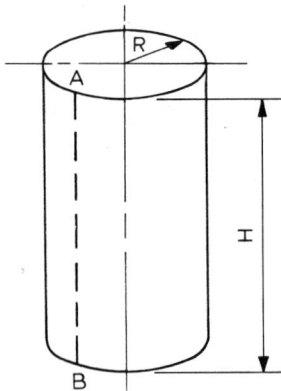

Surface Area

The total surface area is the sum of the areas of the curved surface, plus the areas of the two ends.

Curved surface—imagine a cut made down the outside surface, as AB in the illustration. The surface could then be 'unrolled' to form a rectangle whose area $= 2\pi RH$

$$\text{Area of each end} = \pi R^2$$

$$\text{Total area} = \underline{2\pi RH + 2\pi R^2}.$$

Note: examine the question closely and decide whether two ends, one end or no end is required.

Volume

As in the prism, the volume of a cylinder may be found by multiplying the area of the 'end' by the length of the cylinder.

$$\text{Volume} = \underline{\pi R^2 \times H, \text{ or } \pi R^2 H.}$$

Example—A cylinder is 14 mm diameter and 30 mm high, closed both ends.

$$\text{Surface area} = 2\pi RH + 2\pi R^2$$
$$= 2 \times \frac{22}{7} \times 7 \times 30 +$$
$$2 \times \frac{22}{7} \times 7 \times 7$$
$$= 1\,320 + 308$$
$$= \underline{1\,628 \text{ mm}^2}.$$

$$\text{Volume} = \pi R^2 H$$
$$= \frac{22}{7} \times 7 \times 7 \times 30$$
$$= \underline{4\,620 \text{ mm}^3}.$$

Problems

1. A tube is made of thin sheet metal. It is 35 mm diameter and 60 mm long. What area of sheet is in the finished tube?
   (7 600; 7 000; 6 600; 6 000)
2. A solid piece of material is 28 mm diameter and 40 mm long. Find its total surface area.
   (4 752; 4 900; 5 106; 5 212)
3. Twenty-five discs are to be plated. They are all of the same size, i.e. 7 cm diameter, 0·5 cm thick. What is the total area to be covered?
   (2 500; 2 400; 2 300; 2 200)
4. A container having an open top is 5·6 cm diameter and 12 cm high. Determine its outer area to the nearest cm².
   (316; 260; 236; 200)
5. A closed cylindrical tank, 11·5 m high, has a diameter of 7 m. Find its surface area.
   (350; 330; 310; 300)
6. A steel bar, 4·0 cm diameter is 35 cm long. Find its volume.
   (280; 360; 440; 500)
7. Find the volume of a container which is 40 mm diameter and 70 mm high.
   (100 000; 88 000; 80 000; 77 000)
8. A cylindrical oilcan is 8 cm diameter. It is filled to a depth of 14 cm. What volume of oil is then in the can?
   (750; 608; 664; 704)
9. The internal diameter of a hydraulic cylinder is 7 cm. Determine the volume of oil pumped into the cylinder when a stroke of 40 cm is made.
   (1 540; 1 500; 1 480; 1 400)
10. A drain tray is 0·6 m diameter. What volume of liquid, in cm³, is in the tray when the depth of liquid is 35 mm?
    (9 900; 9 500; 9 000; 8 800)

Answers: **1** *c*; **2** *a*; **3** *d*; **4** *c*; **5** *b*; **6** *c*; **7** *b*; **8** *d*; **9** *a*; **10** *a*.

Mass of Containers and Contents

Mass of Container

Thin Material

First, find the total surface area. Then, multiply this figure by the mass of the material per square unit.

Example—A rectangular tank, open at the top, is 60 cm long, 50 cm wide and 20 cm deep. Find its mass if the metal has a mass of 2 g/cm².

$$\text{Area} = 2 \times 60 \times 20 + 2 \times 50 \times 20 + \\ 60 \times 50$$
$$= 2\,400 + 2\,000 + 3\,000$$
$$= 7\,400 \text{ cm}^2$$
$$\text{Mass} = 7\,400 \times 2$$
$$= 14\,800 \text{ g}$$
$$= \underline{14.8 \text{ kg.}}$$

Thick Material

Find the volume by treating the vessel as solid. Then, subtract the inside volume. This gives the volume of material in the container, and multiplying this by the density of the material we obtain its mass.

Example—A thick cylinder, open at the top, has an internal diameter of 12 cm, and an external diameter of 14 cm. It is 8 cm deep on the outside and 7 cm deep on the inside. The density of the material is 7.5 g/cm³. Find the mass of the cylinder.

$$\text{Outside volume} = \frac{22}{7} \times 7 \times 7 \times 8$$
$$= 1\,232 \text{ cm}^3$$
$$\text{Inside volume} = \frac{22}{7} \times 6 \times 6 \times 7$$
$$= 792 \text{ cm}^3$$
$$\text{Then volume of cylinder} = 1\,232 - 792$$
$$= 440 \text{ cm}^3$$
$$\text{Mass of cylinder} = 440 \times 7.5$$
$$= 3\,300 \text{ g}$$
$$= \underline{3.3 \text{ kg.}}$$

Mass of Contents

First, find the volume of the contents and then multiply by the density.

Example—In question 9, page 6, the volume of the container was 1 540 cm³. If filled with oil, density 0·8 g/cm³, find the mass of the oil.

$$\text{Mass} = 1\,540 \times 0.8$$
$$= \underline{1\,232 \text{ g.}}$$

Problems

1   In a workshop, the container used for soldering flux is in the shape of a cube. The sides are 15 cm long. It is open at the top. Find its mass when the mass of the metal is 0·08 g/cm². (100; 90; 80; 70)

2   A thick plastic container is open at the top. Its inside and outside diameters are 70 and 80 mm, respectively. The inside and outside depths are 60 and 70 mm, respectively. Determine its mass in grammes if the material's density is 1 100 kg/m³. (133; 123; 113; 110)

3   A deep tray, used for freezing, is 10 cm wide, 60 cm deep and 60 cm long. The mass of the metal is 2·5 g/cm². Determine the mass of the tray in kg. (25; 22·5; 20; 17·5)

4   A brass cartridge may be considered to be a closed cylinder, 1·4 cm diameter, 7 cm long. The mass of the brass is 0·5 g/cm². What is the mass of each cartridge, to the nearest gramme? (14; 15; 16; 17)

5   In the example on 'thin material' in the opposite column, find the mass of fluid when the tank is full. The density of the liquid is 900 kg/m³. (65; 60; 56; 54)

Answers: **1** *b*; **2** *a*; **3** *b*; **4** *d*; **5** *d.*

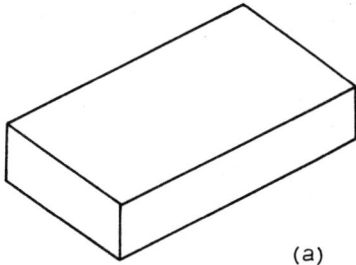

(a)

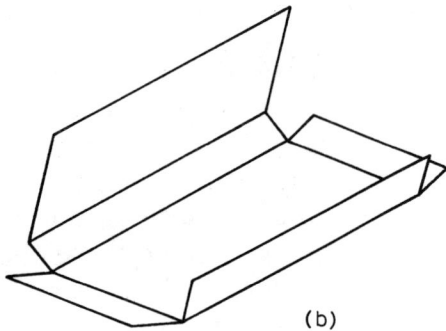

(b)

Size: 100 × 50 × 20 mm

Development of Right Solids

Prisms

The development of a solid means laying out its surface in one plane. A sheet of material cut to this shape could be folded and joined so as to produce the solid shape. In this first example, the solid is illustrated on the left in (a), and the surfaces are opened out a little in (b); the developed shape is in (c), below. Chain lines indicate 'fold lines'. In these exercises no allowance has been made for joints.

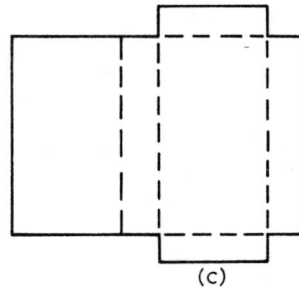

(c)

Cylinders

Illustration (a), left, shows a cylinder closed at both ends. Imagine it to be cut around the top and bottom edges and on a straight line joining top and bottom. Illustration (b) shows it prised open. The curved part, when laid flat, forms a rectangle, as in (c), below.

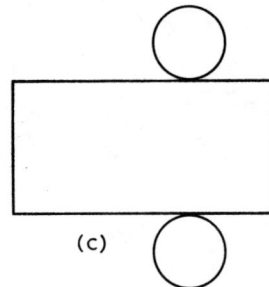

(a)

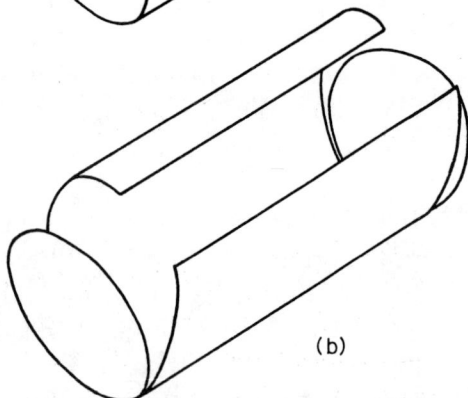

(b)

Size: 50 mm dia. × 120 mm long

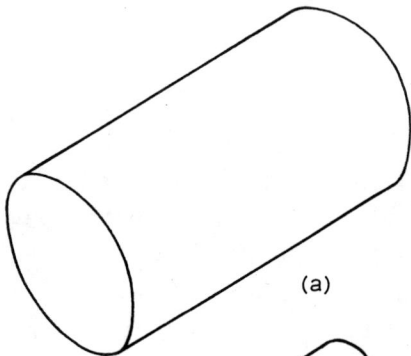

(c)

Pyramids

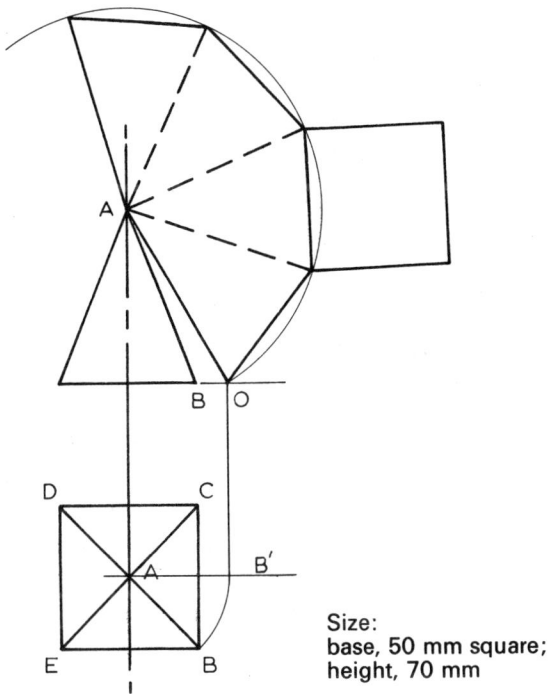

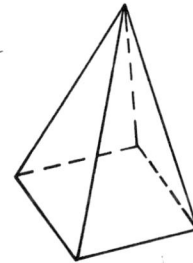

The illustration above is of a square pyramid. To draw the development, first the true length of the sloping edge must be determined, as shown on page 12 of Base book.

On the left, the line AB is swung round to AB^1, projected upward and gives the point O. AO is the required line of true length, on the elevation. On this view, centre A and radius AO are used to give an arc upward and to the right. Step off the length of base four times on the arc. Complete as in the drawing.

Size:
base, 50 mm square;
height, 70 mm

Cones

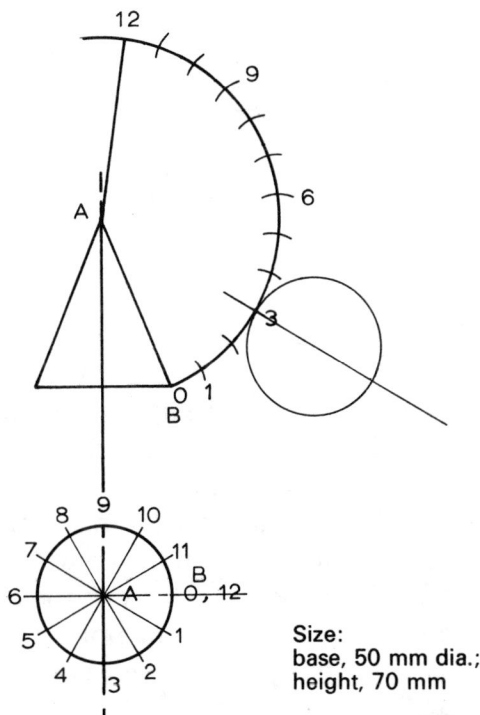

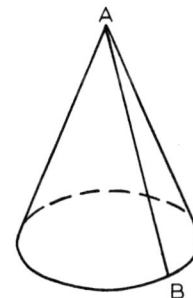

The illustration above shows a cone on which the line AB is a typical line on which to cut the cone and open it out.

The development is shown on the left. The plan, a circle, is divided into 12 sectors from any one of which the chordal length 0–1, 1–2, 2–3, etc., may be used to step out twelve chords along the arc in the elevation swung round, with A as centre and radius AB. The circle on the development may be drawn in any convenient position.

Size:
base, 50 mm dia.;
height, 70 mm

Frustum of a Pyramid

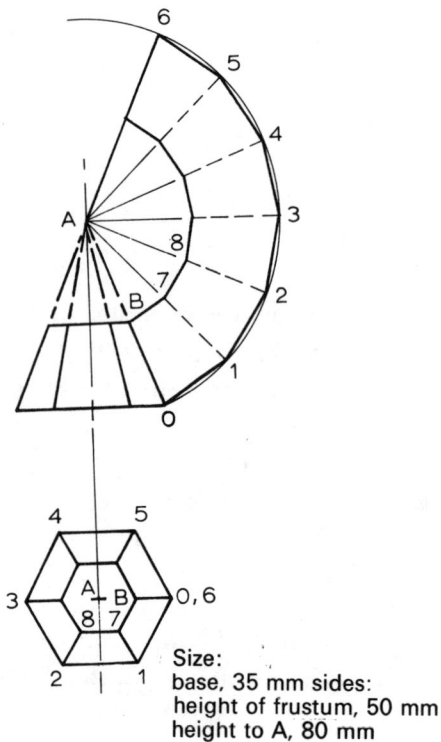

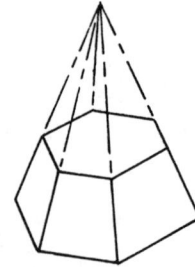

The illustration above is of a hexagonal pyramid with the 'top' cut off, i.e. a frustum of a pyramid.

To develop the sides of it, start as with the pyramid on page 9, except that in this case the true length of a side is already on the figure—BO, left. Centre A, radius AO, draw arc. Centre A, radius AB, draw arc. Step off 0–1, 1–2, etc., and on the smaller arc B–7, 7–8, etc. With straight lines complete the development, as shown on the left.

Size:
base, 35 mm sides:
height of frustum, 50 mm
height to A, 80 mm

Frustum of a Cone

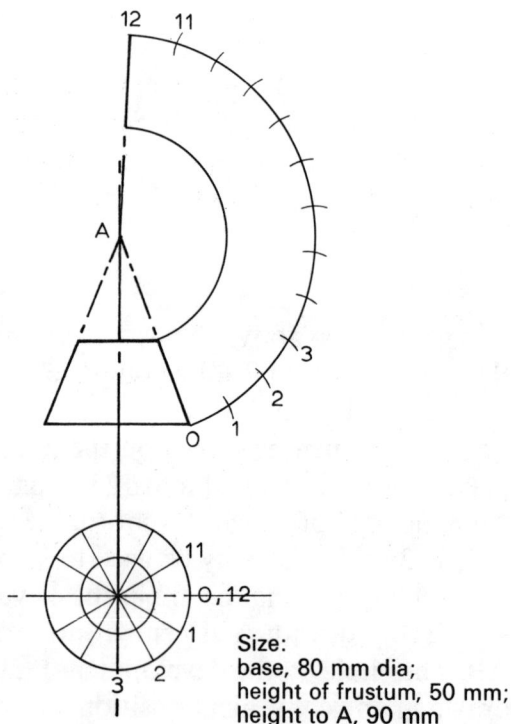

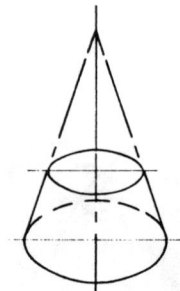

The illustration above shows the frustum of a cone. To the left is the development of the curved surface, which is begun by dividing the circle in the plan at intervals of 30°. Draw the two arcs from centre A.
(*Note:* if only the frustum is given, continue the side lines so as to obtain the apex A.)
Step off 12 chordal lengths as with the cone on page 9. This determines the point 12 on the outer arc. The figure is then easily completed.

Size:
base, 80 mm dia;
height of frustum, 50 mm;
height to A, 90 mm

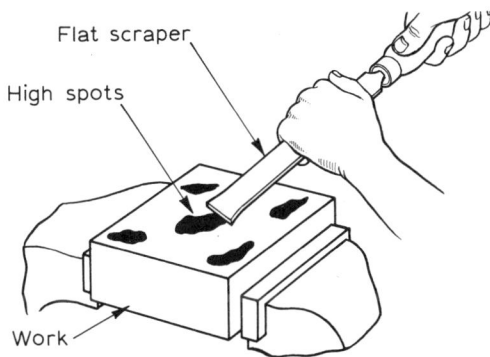

THE FLAT SCRAPER

IRREGULAR HIGH SPOTS

FINISHED SURFACE

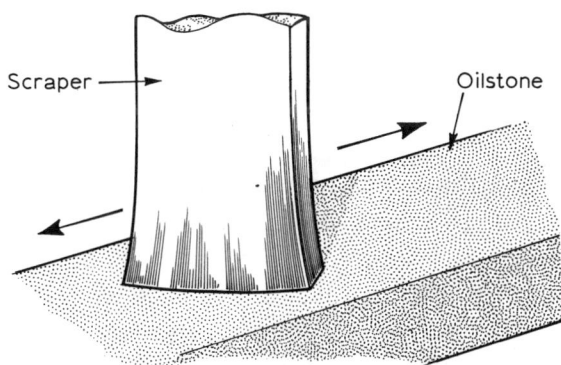

SHARPENING A FLAT SCRAPER

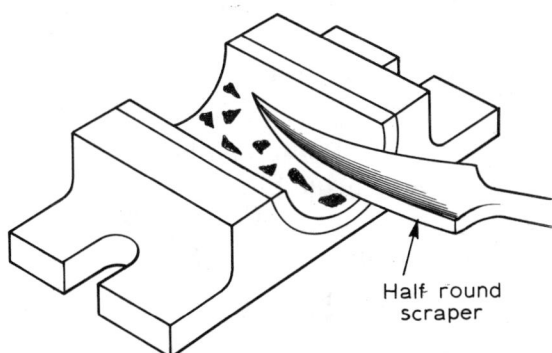

SCRAPING A BEARING

Scrapers

The scraper is used to improve the flatness or in some cases the roundness of component parts. Scrapers must have a very keen cutting edge and must be left in a dead hard condition after heat treatment.

Scraping a Flat Surface

The surface to be scraped should be machined as flat as possible. Engineers' blue is then lightly smeared over an area of a surface plate. After the work has had any burrs removed it is wiped clean. The face to be scraped is then placed carefully face downwards on the blued area of the surface plate and pushed lightly to and fro. The portions of the work which are high will be covered with blue.
The effect can be seen in the left-hand diagram. Scraping then takes place to remove these high spots. This process is repeated until the final result is as shown in the right-hand diagram.

Sharpening a Flat Scraper

To keep the cutting edge keen it must be frequently sharpened during use. An oilstone is used for this. The scraper is sharpened by holding it vertically and moving it across the stone (left).
Hold the scraper at an angle to the edge of the stone to prevent grooves forming. Any burrs on the sides of the scraper can be removed by holding it flat and passing it lightly over the stone.

Half Round Scraper

This is used for scraping bearings. An accurate round bar of the correct diameter is lightly smeared with engineers' blue. It is then rotated in the bearing. The blue portions (high spots) are then removed with the scraper.

B4   Friction and Lubrication   B4

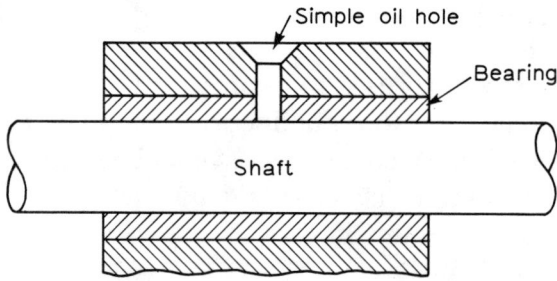

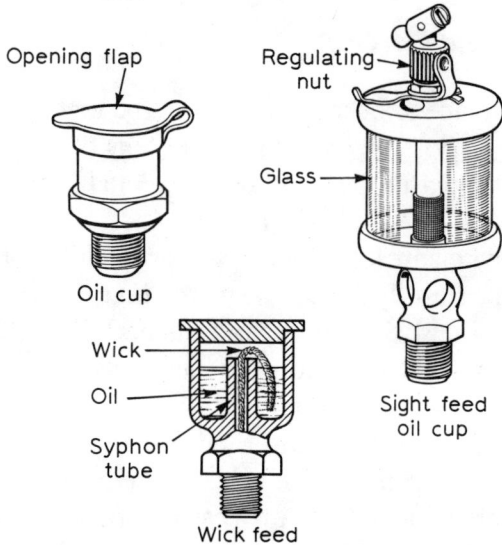

GRAVITY FEED LUBRICATION

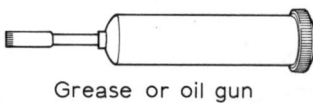

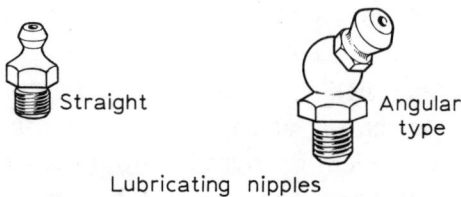

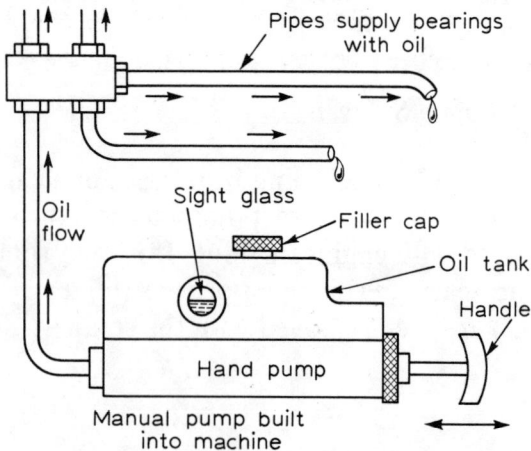

Grease or oil gun

Straight

Angular type

Lubricating nipples

FORCE OR PRESSURE FEED LUBRICATION

Lubrication Systems

A machine tool is an expensive piece of equipment and usually contains many moving parts.

Efficient lubrication plays a vital role in reducing the friction between these moving parts, which will reduce wear.

We have already seen in the Base book how one metal surface will tend to float on the other once a lubricant is introduced.

A good lubricant must be able to maintain this film whilst under load. However the oil must not be too thick or the machine parts would be hard to turn.

Types of Lubricant

Oil is used to lubricate gearboxes, slideways and many moving parts of machine tools.

Grease is used for ball and roller bearings as well as other parts.

Always

Use the correct grade of lubricant or serious damage to the machine may result.

Methods of Application

Lubrication can be divided into three groups:
  (a) Gravity Feed
      Here the lubricant is applied by using a filler or oil can. The oil eventually finding its way to the moving part.
  (b) Force or Pressure Feed
      A grease or oil gun can be used to force the lubricant through grease nipples to the required point. Manual pumps are used to supply various parts of the machine. The oil is held in a tank and when the pump handle is moved oil travels through the copper pipes. Normally one or two strokes per shift will provide sufficient lubrication.

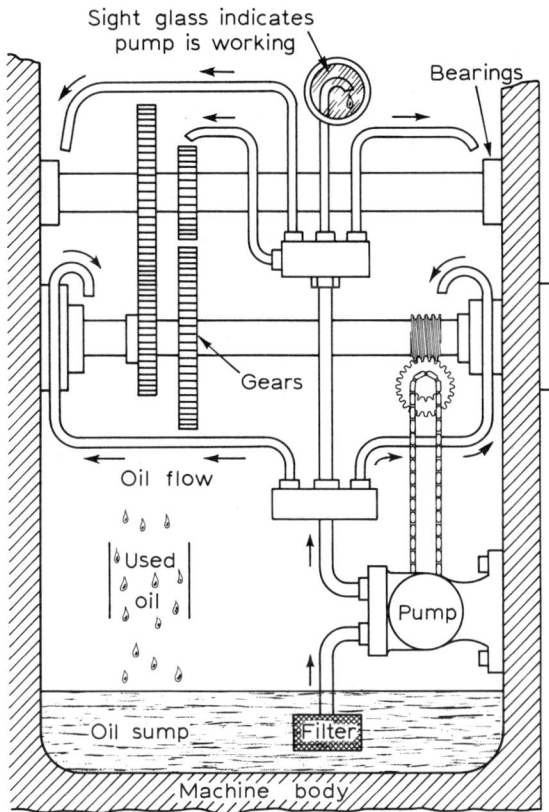

THE OIL PUMP

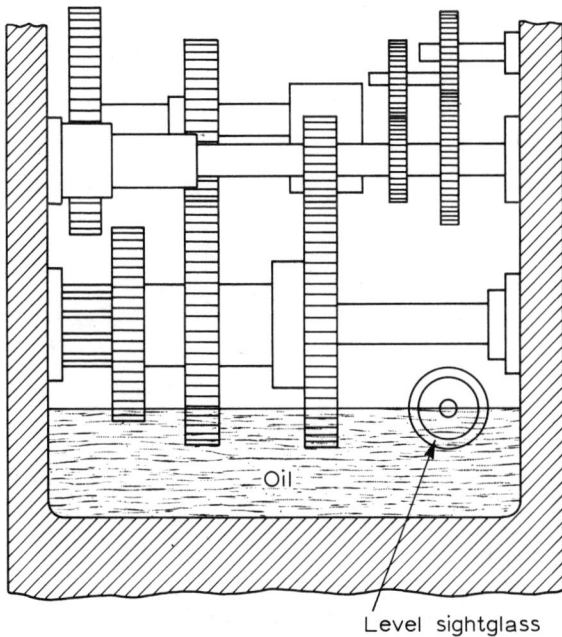

SPLASH LUBRICATION

Lubrication Systems

Force or Pressure Feed

So far we have dealt with the hand pump method of applying a lubricant. The disadvantage of this method is that regular oiling or greasing does not always take place.

Continuous Pressure Feed

A more efficient method of lubrication is the oil pump which is incorporated in and is driven by the machine. The general arrangement of this method is shown opposite.

The oil is contained in a sump. When the pump is working oil is drawn from the sump and passes through the pipes to the respective bearings or gears, etc. It then drains back into the sump and allows for a continuous flow to take place.

Splash Lubrication

This is the third method of lubrication. A simple example is shown here. The gears are shown partly immersed in the oil. On rotating they pick up the oil and carry it to other gears and bearings, etc.

Lubricants and Cleanliness

Lubricating oils and grease should *always* be kept clean. Never allow dirt of any kind to enter containers housing lubricants. Always ensure that the oil can or grease gun is clean before filling with lubricant. Wipe the grease or oil nipples before injecting a lubricant.

B7   Mechanical Methods of Material Removal   B7

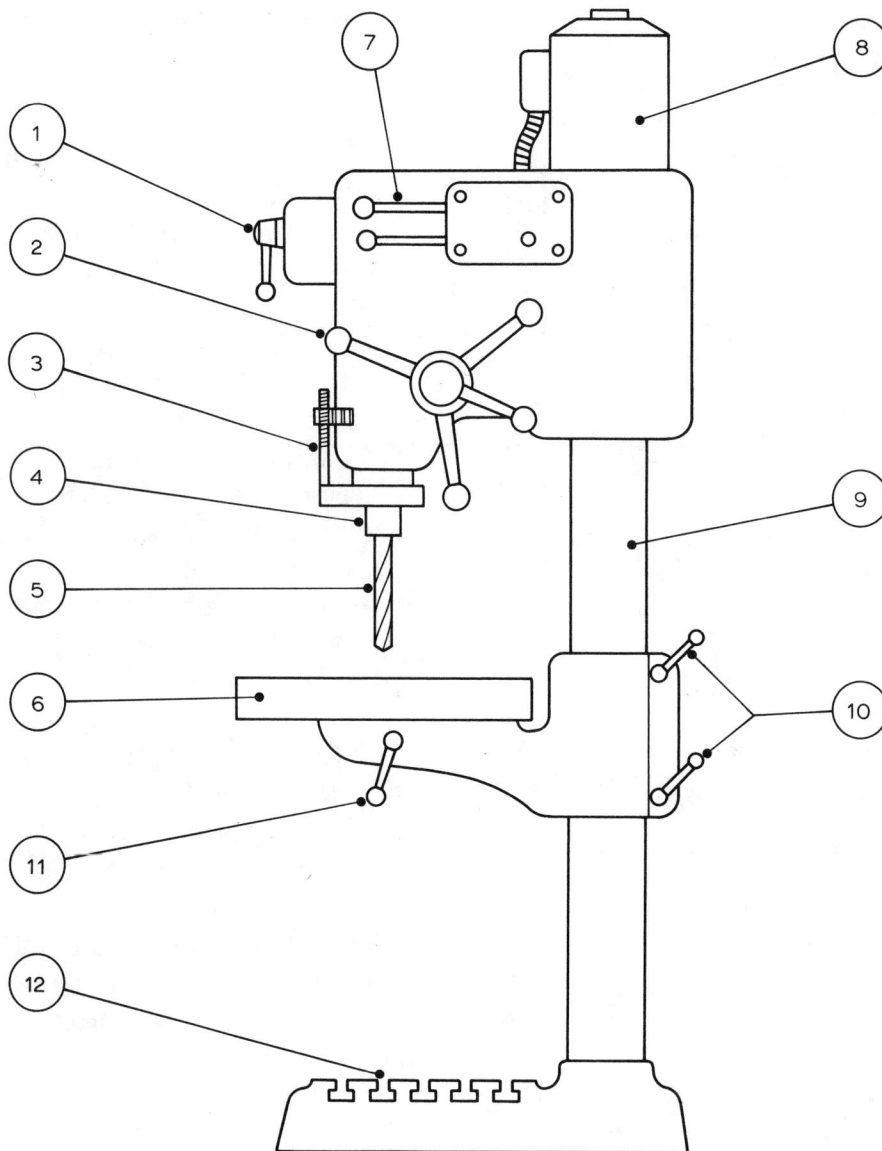

Parts of the Pillar Type Drilling Machine

1   Stop/start switch (electrics).

2   Hand or automatic feed lever.

3   Drill depth stop.

4   Spindle.

5   Drill.

6   Table.

7   Speed change levers.

8   Motor.

9   Pillar.

10   Vertical table lock.

11   Table lock.

12   Base.

Drilling Machines

Holding the Drill

Drills with parallel shanks are held in a chuck (left).
The shank must be in good condition if the drill is to revolve accurately.
The chuck is usually located in the machine spindle by means of a tapered arbor fitted to the top of the chuck.
Never allow parallel-shank drills to twist in the chuck. This will affect the true running of the drill and will also obliterate the size of drill which is indicated on the shank.

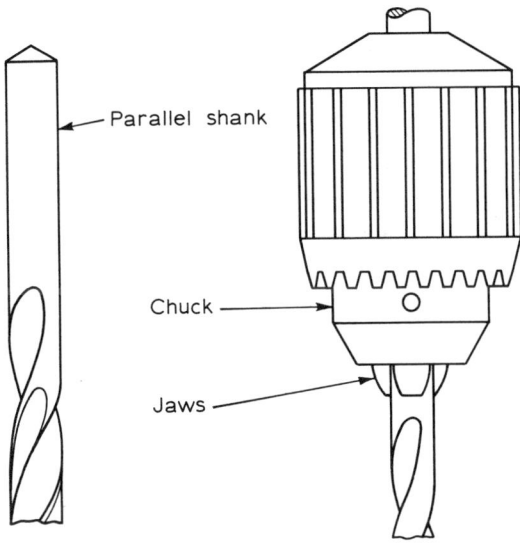

HOLDING A PARALLEL SHANK DRILL

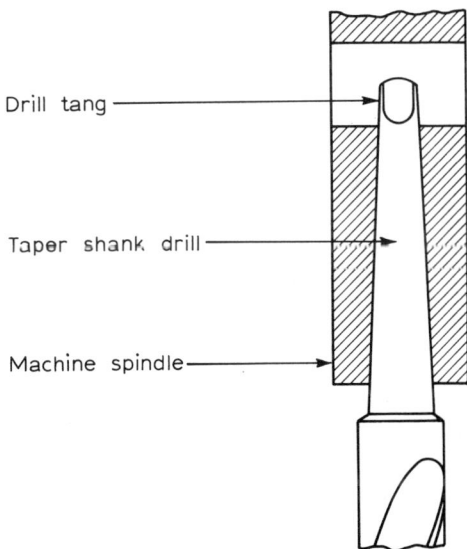

Parallel shank

Chuck

Jaws

Holding Taper Shank Drills

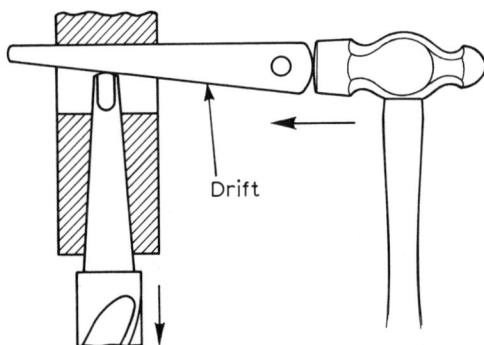

Some drills have taper shanks. The shank is located in the taper in the machine spindle. To ensure true running and a positive grip, the mating portions of the drill and machine spindle should be thoroughly cleaned before inserting the drill.
The drill is forced into the taper by a light tap with a soft hammer or by using the handwheel and lightly forcing the drill down on to a piece of wood placed on the machine table.

Drill tang

Taper shank drill

Machine spindle

HOLDING TAPER SHANK DRILLS

To extract a taper shank drill a taper drift is used. Do not allow the drill to fall when extracting, damage to the cutting edge of the drill or to the machine table may result.

Drift

EXTRACTING TAPER SHANK DRILLS

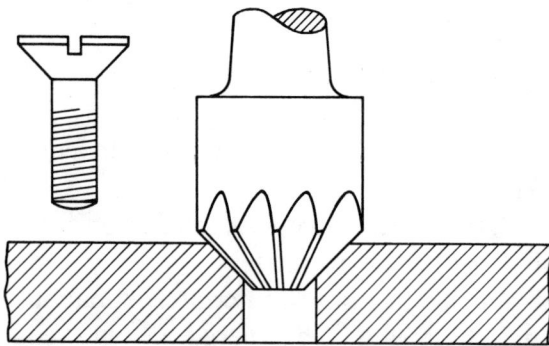

COUNTERSINKING

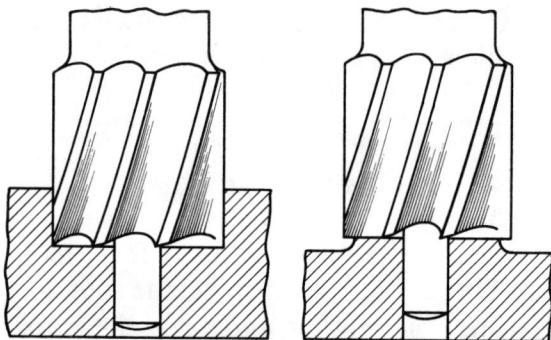

COUNTERBORING   SPOTFACING

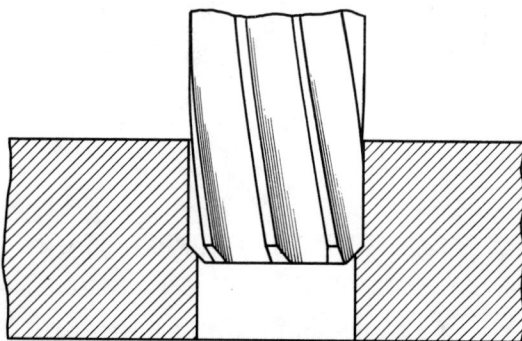

REAMING

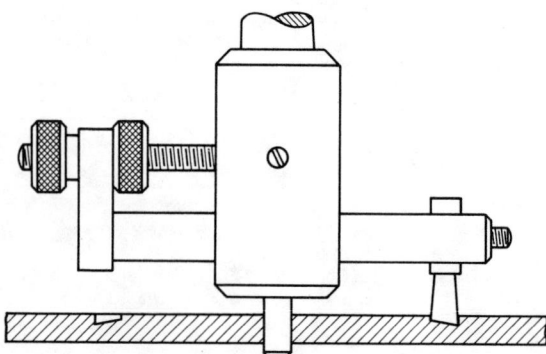

HOLE CUTTING

Drilling Machines

Countersinking

A countersink is used to countersink a hole for rivets or screws. Countersinks are usually supplied with an included angle of 60° or 90°.

Counterboring

Counterboring tools are used to enable screws or bolts to lie below the surface of the work. The pilot acts as a guide and prevents deflection of the tool during cutting.

Spotfacing

The spotfacing tool is similar to the counterboring tool. It is used to true up the surface of the work around a hole. Bolts or nuts will then seat squarely in position.

Reaming

A reamer is used to finish off a drilled hole to size. It has several cutting edges and produces a very smooth accurate hole.

Hole Cutting

Hole-cutters are mainly used for cutting large diameter holes in sheet metal. The cutting tool is removable for sharpening.

The Centre Lathe

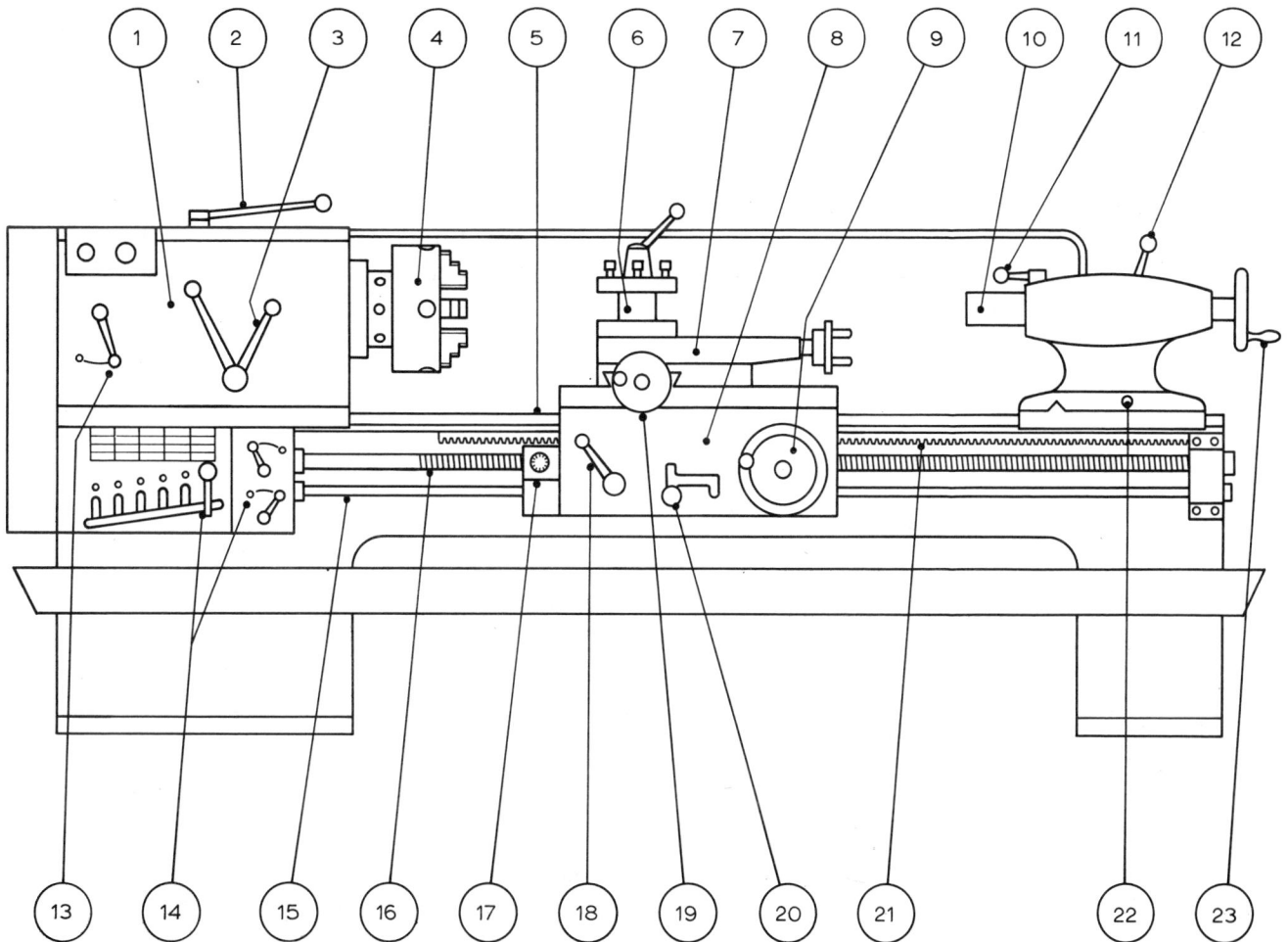

Parts of the Centre Lathe

1	All-geared headstock.	13	Feed reverse lever.
2	Clutch lever.	14	Quick-change gearbox.
3	Speed change levers.	15	Feedshaft.
4	Chuck.	16	Leadscrew.
5	Bed slideways.	17	Screwcutting dial.
6	Four-way turret.	18	Screwcutting lever.
7	Compound slide.	19	Cross-slide handwheel.
8	Apron.	20	Feed engage lever.
9	Sliding handwheel.	21	Rack.
10	Tailstock sleeve.	22	Tailstock offset screw.
11	Sleeve locking lever.	23	Tailstock handwheel.
12	Tailstock clamping lever.		

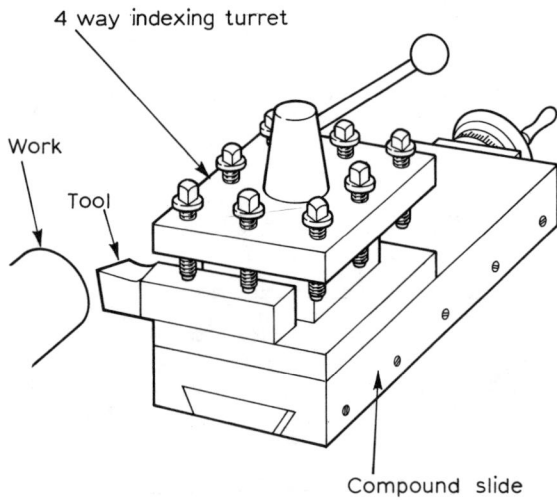

4 way indexing turret

Work

Tool

Compound slide

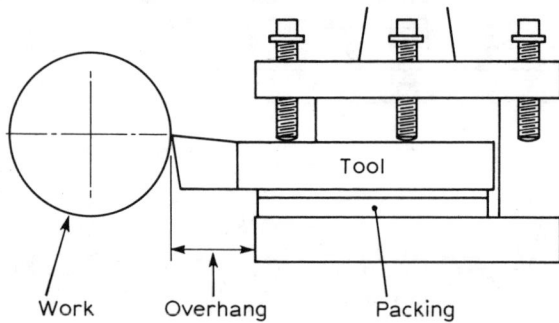

Work   Overhang   Packing

A GOOD TOOL SETUP

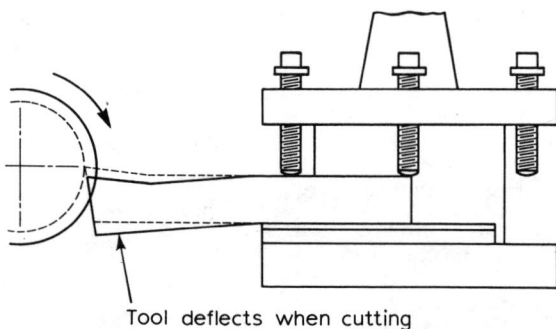

Tool deflects when cutting

TOO MUCH OVERHANG

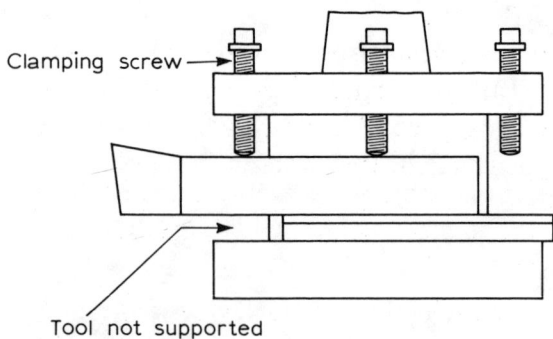

Clamping screw

Tool not supported

TOOL PACKING INCORRECTLY POSITIONED

The Centre Lathe

Holding the Cutting Tool

The four-way indexing turret (left) allows four cutting tools to be held at the same time. This enables a sequence of cutting operations to be carried out without changing tools. As a tool completes its cutting operation the turret is turned through 90° to allow the next tool to come into use.

Setting the Tool

The tool point should always be set on the centre line of the work. For efficient cutting the rigidity of the cutting tool is very important. Always ensure that the tool overhang is as small as possible. Note also the position of the packing pieces.

Excessive tool overhang will cause the tool to deflect when cutting. Heavy cuts may also tend to move the tool in the turret. This will, of course, be due to the large amount of leverage between the tip of the tool and the clamping screws.

Incorrectly Placed Packing

As shown here will also mean that the tool overhang will be excessive. There is also the added danger of bending or even breaking the cutting tool when tightening the front clamping screw.

Packing

When packing is used remember it is better to keep the number of pieces down to a minimum to obtain greater rigidity. One piece of packing is better than a number of pieces amounting to the same size.

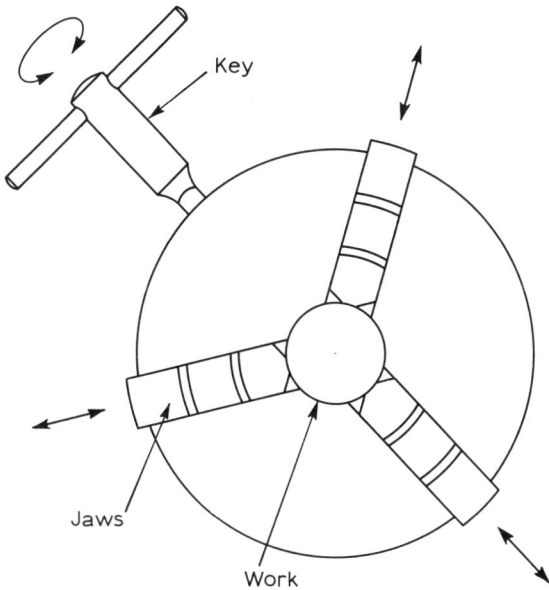

THREE JAW SELF CENTRING CHUCK

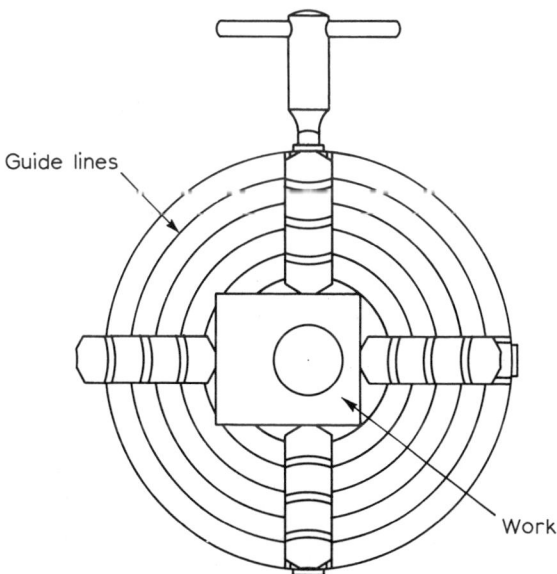

FOUR JAW INDEPENDENT CHUCK

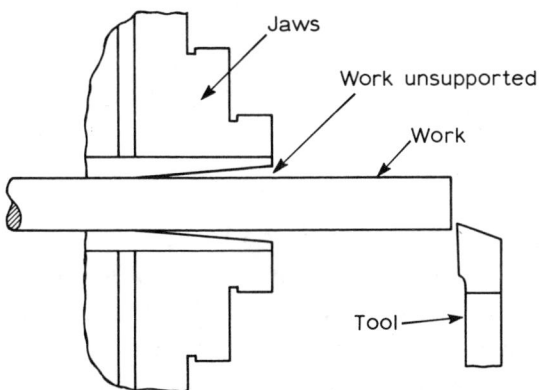

WORK DEFLECTION DUE TO BADLY
WORN CHUCK JAWS

The Centre Lathe

Holding the Work

The three jaw self-centring chuck is used for
gripping round or hexagonal work. As the
key is rotated the jaws will open or close
simultaneously (together).

Four Jaw Independent Chuck

This is used for gripping square or irregular
shaped work. Each jaw is operated inde-
pendently (one at a time). Due to the nature
of its construction this chuck is very rigid
and grips work well.

Balance

Care should be taken when holding any work
that is liable to affect the balance of the
chuck. Very few of the smaller type four jaw
chucks have suitable tee slots to which
balance weights can be attached.
If the chuck cannot be balanced care should
be taken to keep the spindle speed low enough
to avoid vibration.

Worn Chuck Jaws

Chuck jaws can after considerable use become
worn. Wear usually takes place at the front
end of the jaws. The work will then be
gripped only at the back of the jaws. This
means that deflection will take place when
cutting due to the length of unsupported
work.
Worn chuck jaws can be reground by using
a special grinding machine attached to the
lathe.

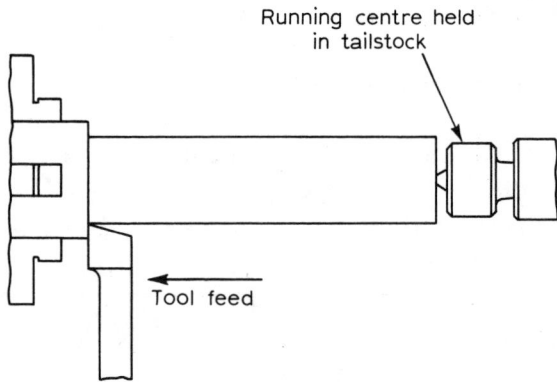

PARALLEL TURNING

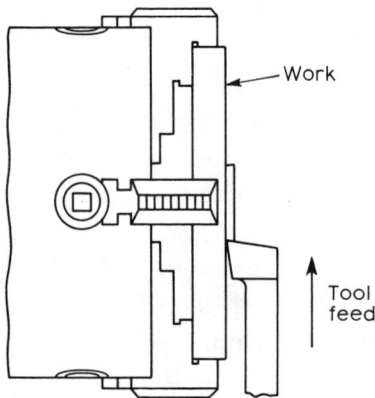

FACING

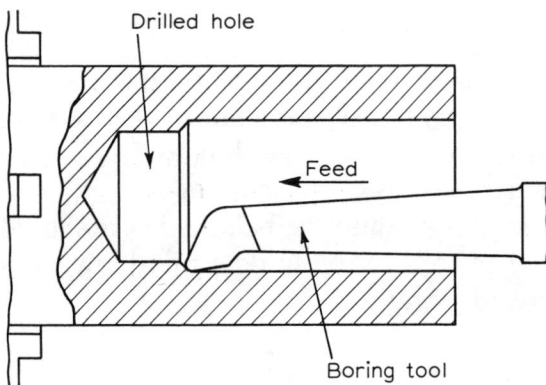

BORING

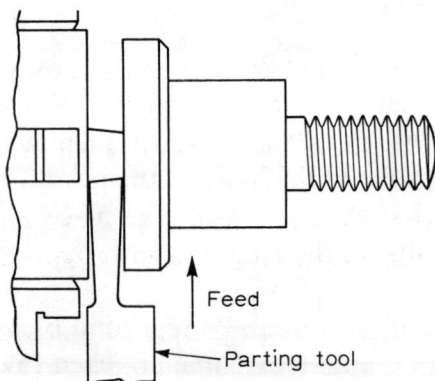

PARTING

The Centre Lathe

Parallel Turning

This diagram shows the direction of tool feed
for parallel turning.
This is sometimes called a sliding cut. Cutting
should not be attempted on work which is
protruding a long way from the chuck unless
it is supported by a centre. Other means of
support are described later.
Remember, to achieve accuracy and a good
surface finish rigidity of the work is important.
Unsupported work will deflect during cutting
and will result in the work not being turned
parallel.

Facing

This shows the direction of tool feed for a
facing cut. This is sometimes called surfacing.
It is good practice to lock the lathe carriage
during cutting. Failure to do this may result
in the carriage being forced away from the
work due to the pressure of the cut.

Boring

It is sometimes necessary to enlarge drilled
holes. A boring tool is used for this purpose.
The cut of course is a sliding one. Boring
tools should be as rigid as possible to avoid
tool deflection. This is very important when
long holes are to be bored. Always make sure
there is sufficient tool clearance to avoid
rubbing and scoring of the work.

Parting Off

This means the cutting off of the workpiece.
Always part off as close to the chuck as
possible. Make sure the tool clearances are
adequate but not too great. Excessive
clearances will only weaken the tool. Avoid
too much overhang. Lock the saddle when
cutting and maintain a constant feed.

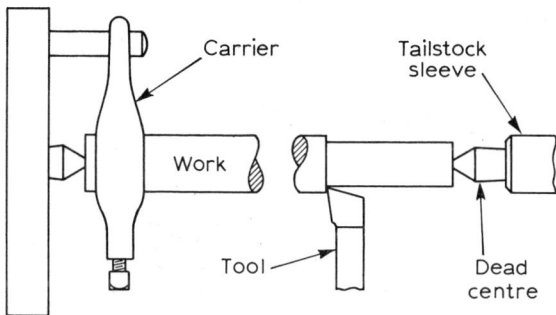

TURNING BETWEEN CENTRES

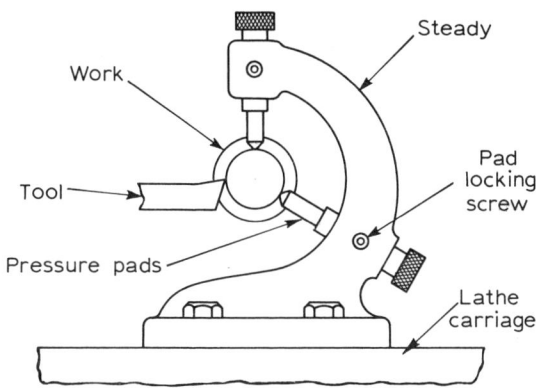

THE TRAVELLING STEADY

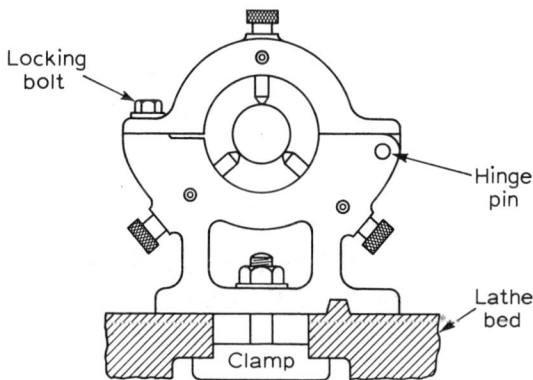

THE FIXED STEADY

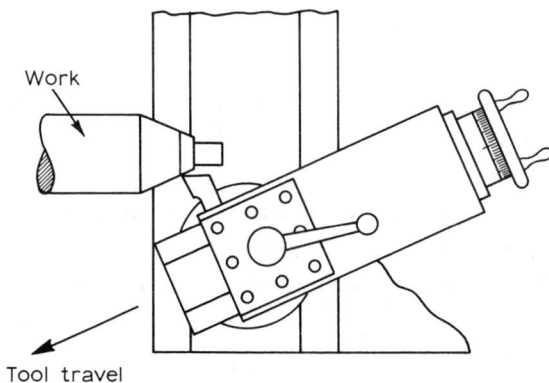

TURNING A TAPER USING THE
COMPOUND SLIDE

The Centre Lathe

Turning Between Centres

The work is first faced and centre drilled each end. A carrier is then attached to one end of the work, and is used to enable the work to be rotated by the driving-plate pin. Grease should of course be placed in the centre drilled hole if a dead centre is to be used. This will reduce friction between the work and the centre. Long workpieces can be supported by using a travelling steady.

Travelling Steady

This is used to prevent the work deflecting during cutting. They normally have two pressure pads which are adjusted until they contact the work just behind the cutting tool. The steady is bolted to the carriage and therefore follows the tool during cutting.

Fixed Steady

Also prevents deflection of the work during cutting. They have three adjustable pressure pads. This type of steady is bolted to the lathe bed and therefore remains stationary. It can be used to support the end of long work protruding from a chuck or to support work between centres, provided the tool movement is not hindered.

Turning Tapers

One common method of turning tapers of short length is to use the compound slide. The slide is set to the required angle and the tool fed by using the handwheel. Other methods of turning tapers use form tools, offset tailstock and taper turning attachment.

21

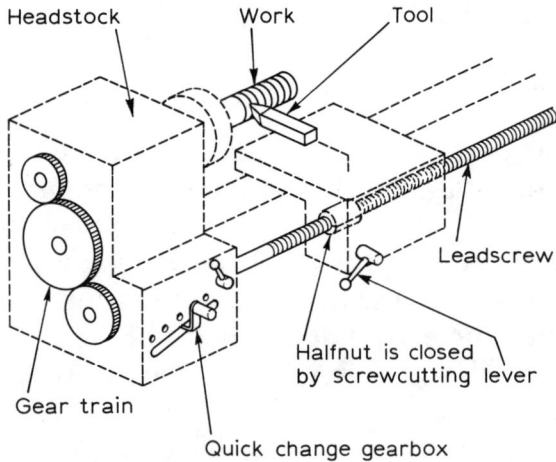

The Centre Lathe

Screwcutting

Cutting screwthreads on a centre lathe is known as screwcutting. A train of gears couples the headstock spindle through to the quick-change gearbox. At this point the correct pitch of thread to be cut can be rapidly selected, the leadscrew will then revolve at the required rate. The halfnut is closed when screwcutting. This means that as the leadscrew revolves the carriage will be driven along by the rotation of the leadscrew (left). If the required pitch is not available on the quick-change gearbox it will be necessary to calculate and then change the gears between the headstock and the quick-change gearbox.

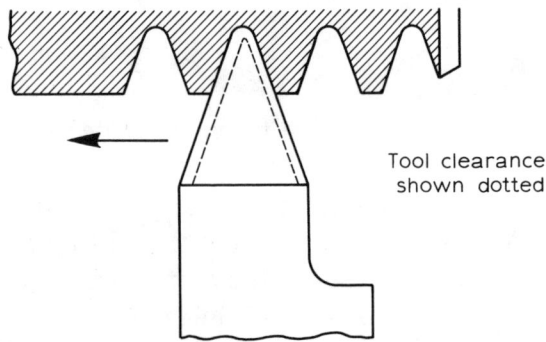

SINGLE POINT SCREWCUTTING TOOL

The Screwcutting Tool

The single-point tool is normally used for screwcutting, although accurately ground tools of full thread form are available. The single-point tool must also be very accurate if a good thread is to be produced.

A screwcutting gauge (left) can be used to check the tool during grinding. Make sure also that the side clearance is adequate for the thread to be cut.

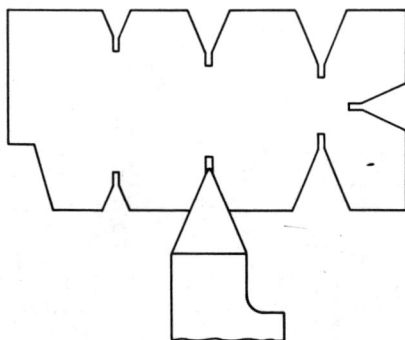

CHECKING THE TOOL ANGLE

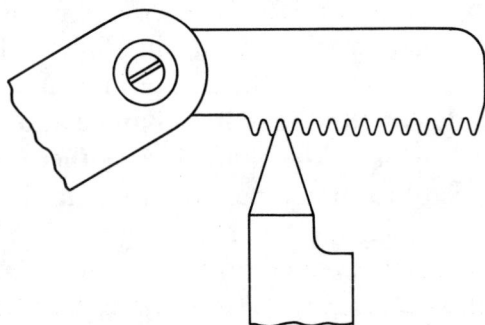

The radius on the tip of the tool will of course produce the root radius on the thread being cut. This radius must be carefully ground or stoned on the tool. It can be checked by using a screwthread pitch gauge. This radius must never be exceeded or a slack thread may result.

CHECKING THE TOOL RADIUS

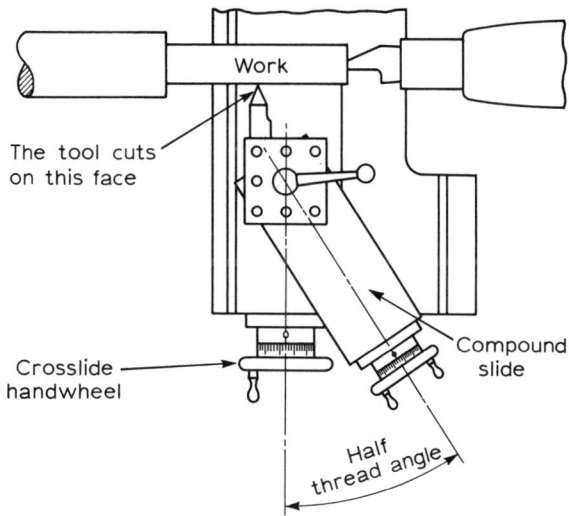

SCREWCUTTING SETUP

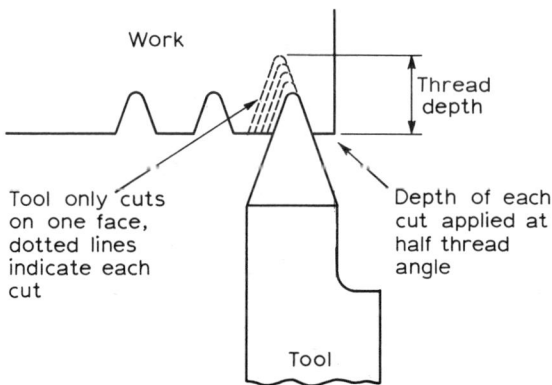

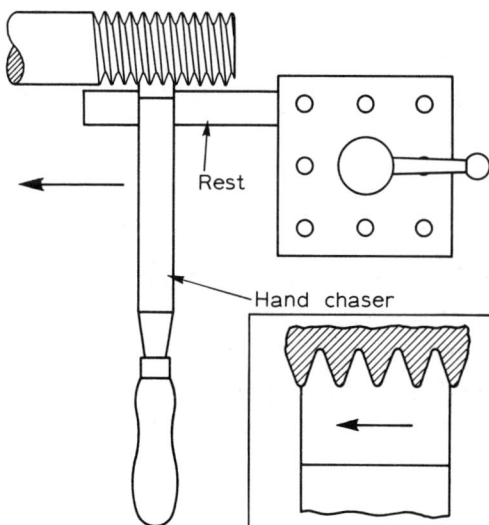

USING A HAND CHASER

The Centre Lathe

The Screwcutting Operation

A screwcutting set-up is shown at the left. It can be seen that the compound slide is set to half the thread angle. The reason for this can be seen in the second diagram. The tool depth for each cut is applied to the compound slide handwheel.

This means that cutting will take place on the leading edge of the tool only. Deflection of the tool is therefore reduced, which would not be the case if the depth of cut were applied to the cross-slide handwheel. In this case cutting would take place on both sides of the tool and may result in scoring of the work, especially when cutting threads of a coarse pitch.

Finishing the Thread

Some threads have a crest radius. It is not possible to finish this with the single-point screwcutting tool. One method of applying this radius is by using a hand chaser. The use of the hand chaser is shown opposite. It can be seen that the body has teeth cut on one end. These teeth conform to the form of the thread being cut. The other end has a tang on which a handle is fitted. The chaser is held in the hand during use and is supported by a rest clamped in the turret. A chaser must be held square to the work or the thread may be ruined.

Safety

Hand chasers can be dangerous if not used with great care. Always make sure that the rest is close to the work and that clothing does not come into contact with moving parts.

THE SHAPING MACHINE

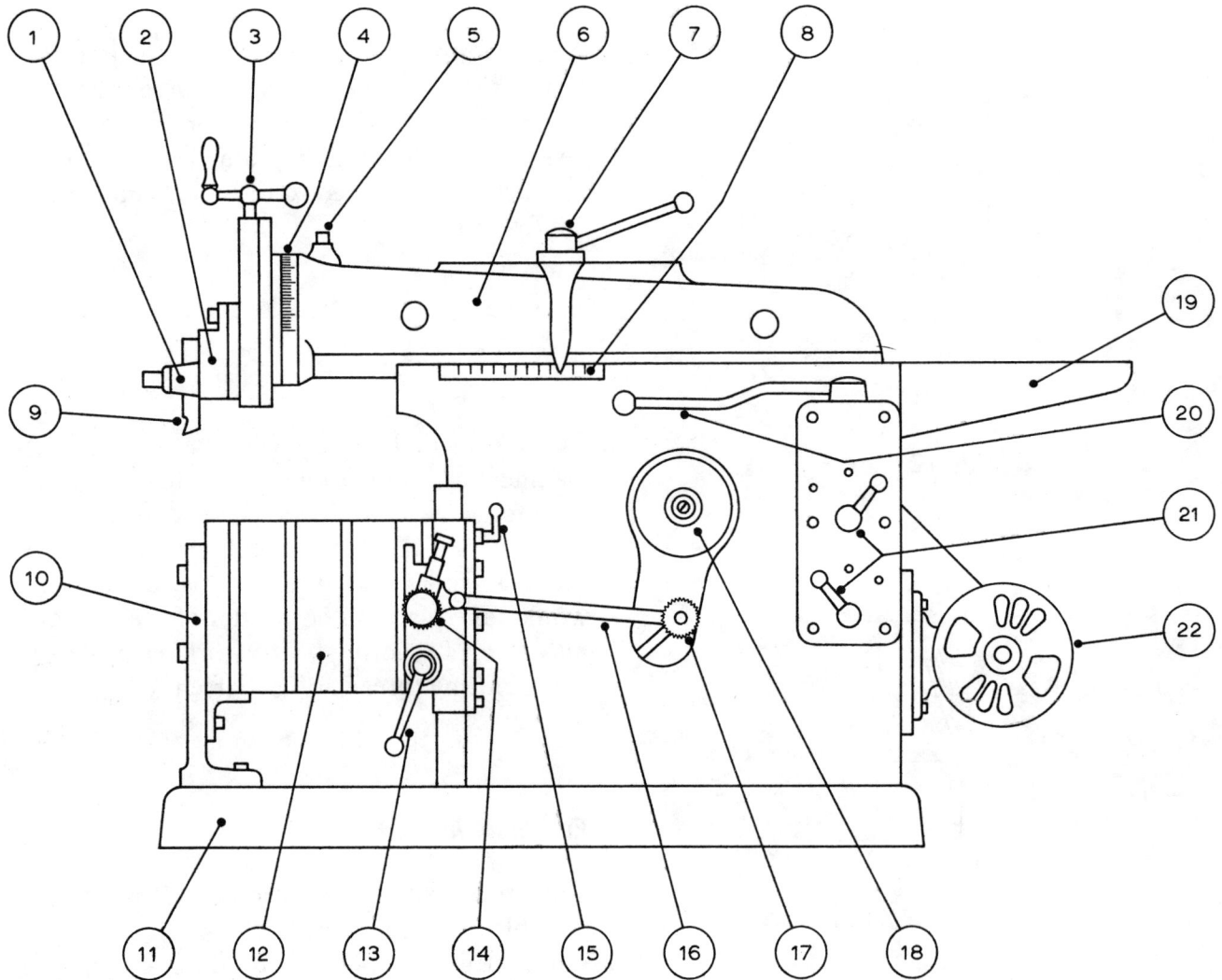

Parts of the Shaping Machine

1	Toolholder.	12	Table.
2	Clapper box.	13	Handle for vertical table movement.
3	Slide handwheel.	14	Horizontal feed ratchet.
4	Head swivels at this point.	15	Table clamp (vertical).
5	Locking screw for swivel head.	16	Feed connecting rod.
6	Ram.	17	Feed adjusting knob.
7	Ram clamping handle.	18	Stroke adjusting spindle.
8	Length of stroke indicator.	19	Ram guard.
9	Cutting tool.	20	Clutch lever.
10	Table support bracket.	21	Speed change levers.
11	Base.	22	Electric motor.

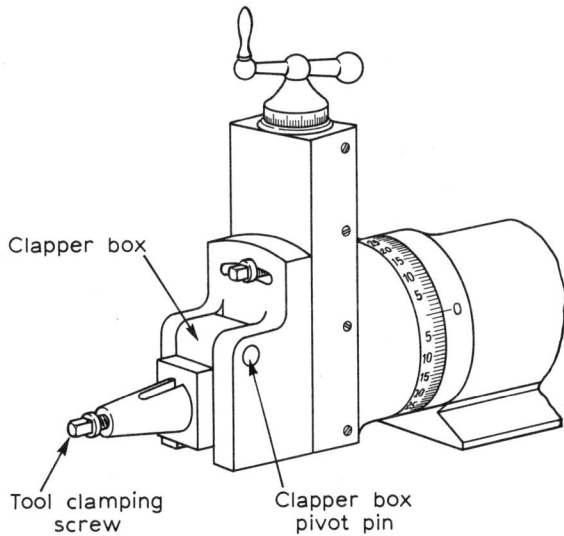

TOOL BOX AND HEADSLIDE

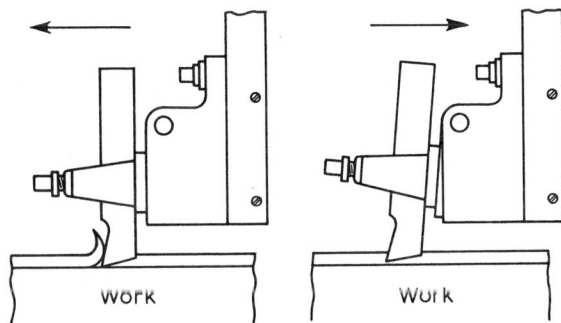

CUTTING STROKE   RETURN STROKE

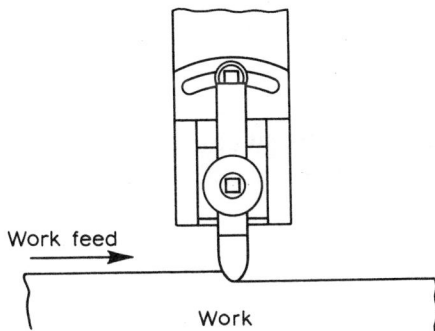

SHAPING A HORIZONTAL SURFACE

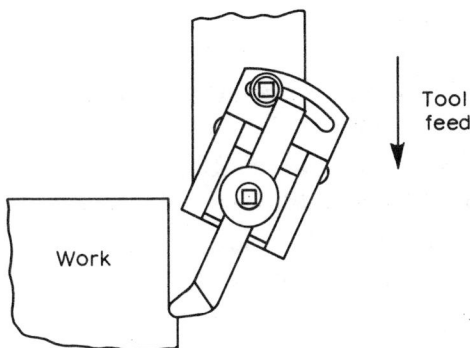

SHAPING A VERTICAL SURFACE

The Shaping Machine

Tool Box and Headslide

The tool is held in a tool post by means of a single screw. The tool post is mounted on a clapper box which pivots on a pin. It pivots to allow the tool to lift clear of the work as it is fed over during the return stroke of the ram. This can be seen in the second illustration.

Slide Handwheel

The slide handwheel controls the depth of cut. It is also used when cutting vertical or angular faces. Normally the tool slide is hand fed. Some machines, however, have a vertical table feed.

Swivel Head

The swivel head enables the tool to be fed at the required angle. When using the swivel head make sure that it clears the body of the machine if long strokes are to be taken.

The Shaping Process

Care must be taken to ensure that the work is fed against the correct side of the tool. The length of stroke must be sufficient to allow the clapper box to drop back into the vertical position before commencing the cutting stroke. Never set the machine to a longer stroke than is necessary. Remember, it costs money to cut fresh air.

Shaping a Vertical Surface

Here the clapper box is swivelled away from the face of the work to be cut. On the return stroke the tool swings away from the work and gives clearance, which prevents the work being scored. The tool feed is, of course, applied by the slide handwheel during or at the end of the return stroke.

25

Dimensions in millimetres

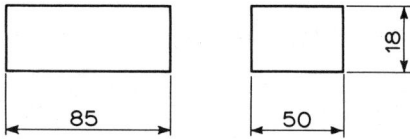

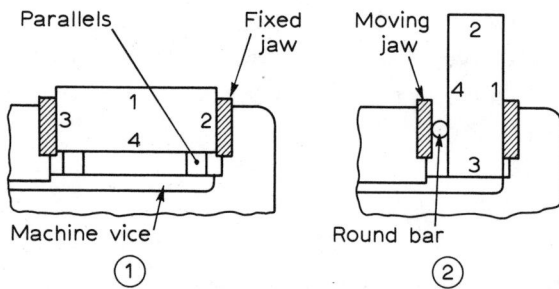

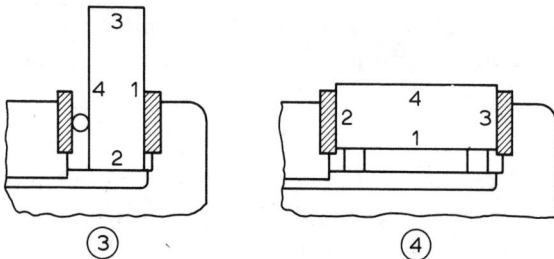

Parallels   Fixed jaw   Moving jaw

Machine vice   Round bar

① ②

③ ④

SHAPING A BLOCK SQUARE

Head set to required angle

Tool feed

Work

SHAPING ANGULAR FACES

Tool feed

Work

SHAPING GROOVES

Shaping a Block Square

1 Place block on parallels in machine vice. Shape first face (wide surface).
2 Place first machined face against fixed jaw and round bar between moving jaw and work. Shape second face.
3 Place second machined face against vice base and round bar between moving jaw and work. Shape third face to required length.
4 Place first machined face on parallels and shape fourth face to required thickness.

It can be seen that the face of the fixed vice jaw governs the squareness of the finished work. To achieve this squareness the fixed jaw must be square to the machine table. The round bar placed between the moving jaw and the workpiece will ensure that the work will be forced squarely against the fixed jaw. When using the round bar, avoid taking heavy cuts. Always make sure that the workpiece is well down on the parallels before taking a cut.

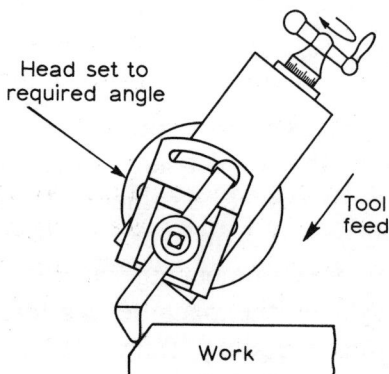

Shaping Angular Faces

The swivel head can be seen set to the required angle. Note the position of the clapper box; this is swung away from the face to be machined. The tool feed is applied by the slide handwheel.

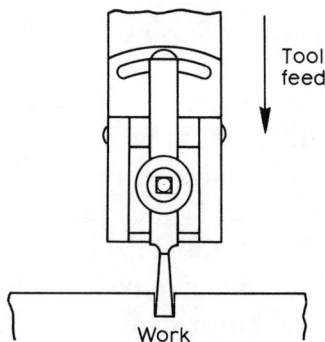

Shaping Grooves

The tool used is similar to a lathe parting tool with the exception of the front face which, of course, must be square. Rigidity of the tool is very important. Avoid excessive overhang and ensure that the tool clearances are not so great as to weaken the tool.

SLOTTING MACHINE

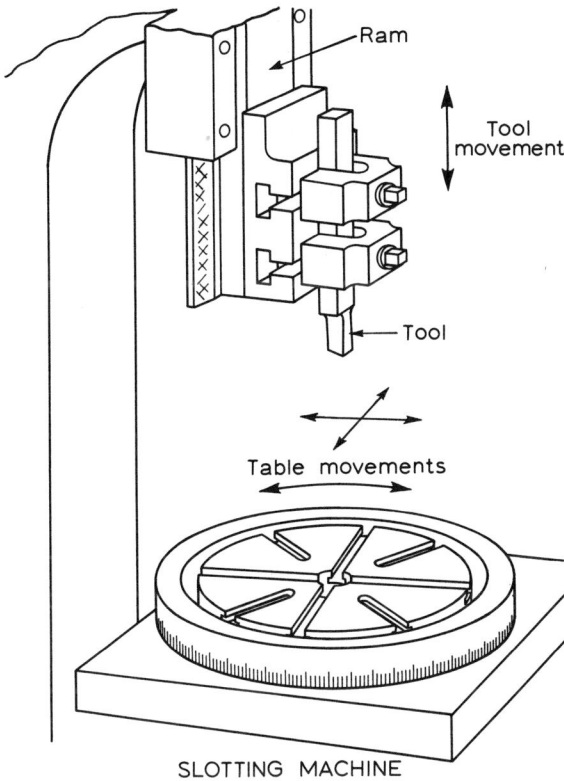

The Slotting Machine

The slotting machine could be considered as a vertical shaper because the ram operates in a vertical plane. Special clamps which locate in tee slots hold the tool vertically. The tool should be robust enough to withstand the cutting forces that are, of course, applied parallel to the shank.

The worktable is of the circular type and is graduated for angular work or where divisions of a circle are required. The circular table is also very useful for machining work of circular shape. This table is mounted on a saddle which may be automatically traversed in the same manner as a milling machine.

Milling machines of the horizontal type can be fitted with slotting heads which have a limited stroke. A circular table can be bolted to the milling machine table if required.

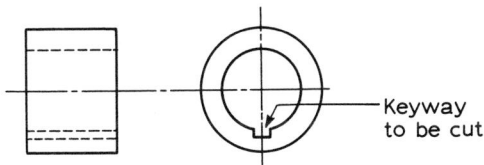

Slotting a Keyway

The tool used is generally smaller in width than the slot to be cut. This enables the slot to be finished accurately by moving the table over the required amount.

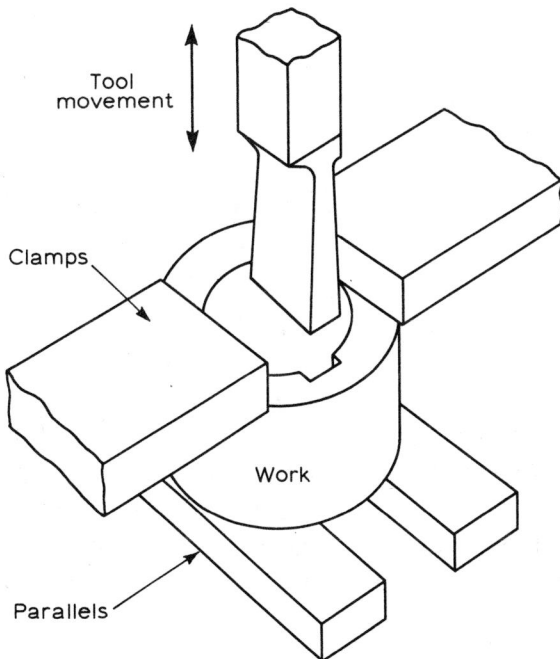

The work should be mounted on parallels in order to give the tool clearance as it passes through the work. Always make sure that the parallels support the work close to the cutting area.

SLOTTING A KEYWAY

THE 'HORIZONTAL' MILLING MACHINE

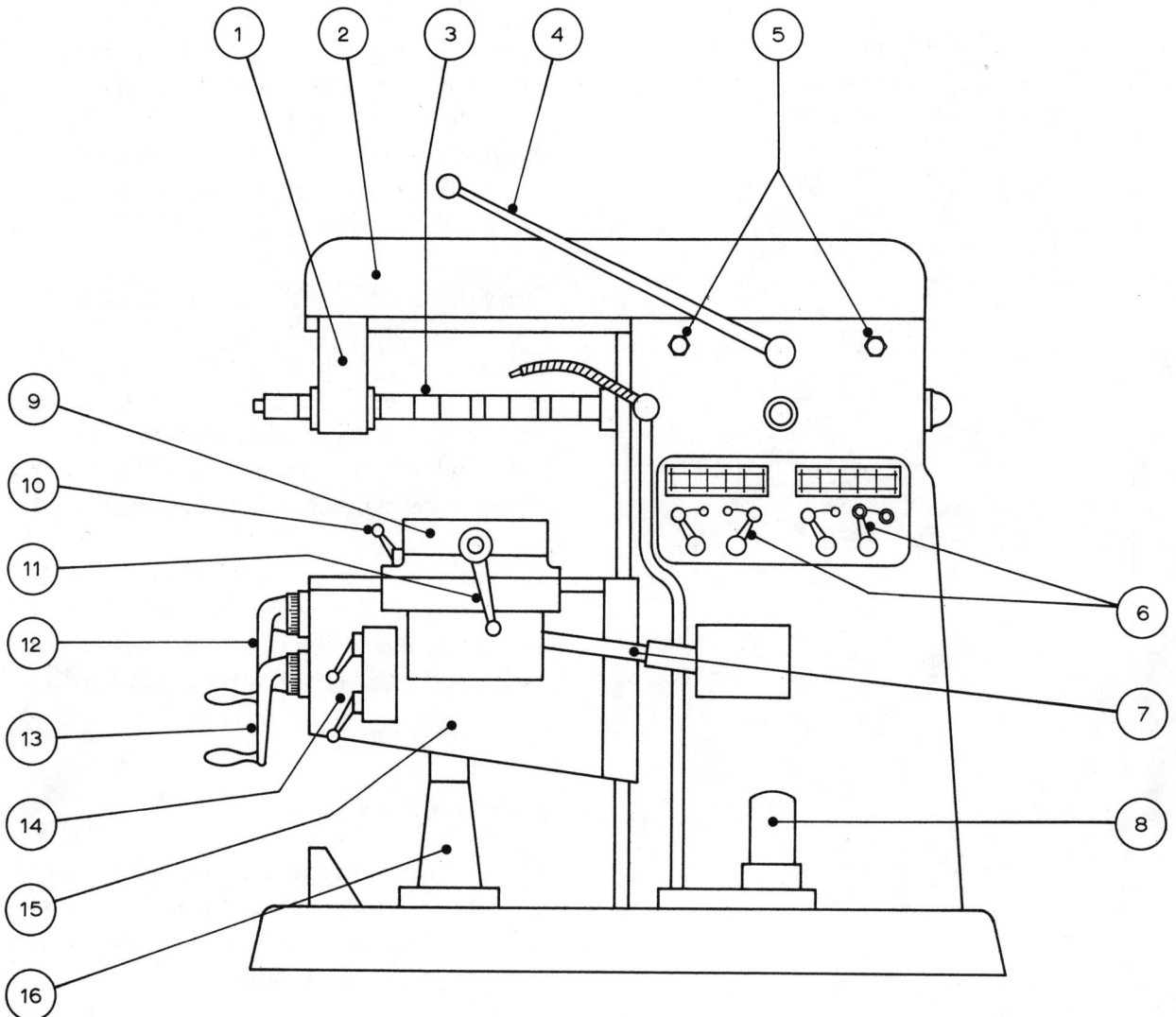

Parts of the Horizontal Milling Machine

1 Arbor support or yoke (which slides on overarm).
2 Overarm.
3 Arbor (on which the cutter is mounted).
4 Clutch lever.
5 Overarm clamping screws.
6 Speed and feed selector levers.
7 Telescopic feedshaft.
8 Coolant pump.
9 Table which contains tee slots for clamping work or a vice.
10 Automatic table feed (longitudinal).
11 Longitudinal feed (hand).
12 Vertical feed (hand) which raises or lowers knee assembly.
13 Cross or transverse handle (hand).
14 Vertical and horizontal automatic-feed levers.
15 Knee which supports table and moves up and down on dovetail slides.
16 Knee elevating screw.

THE 'VERTICAL' MILLING MACHINE

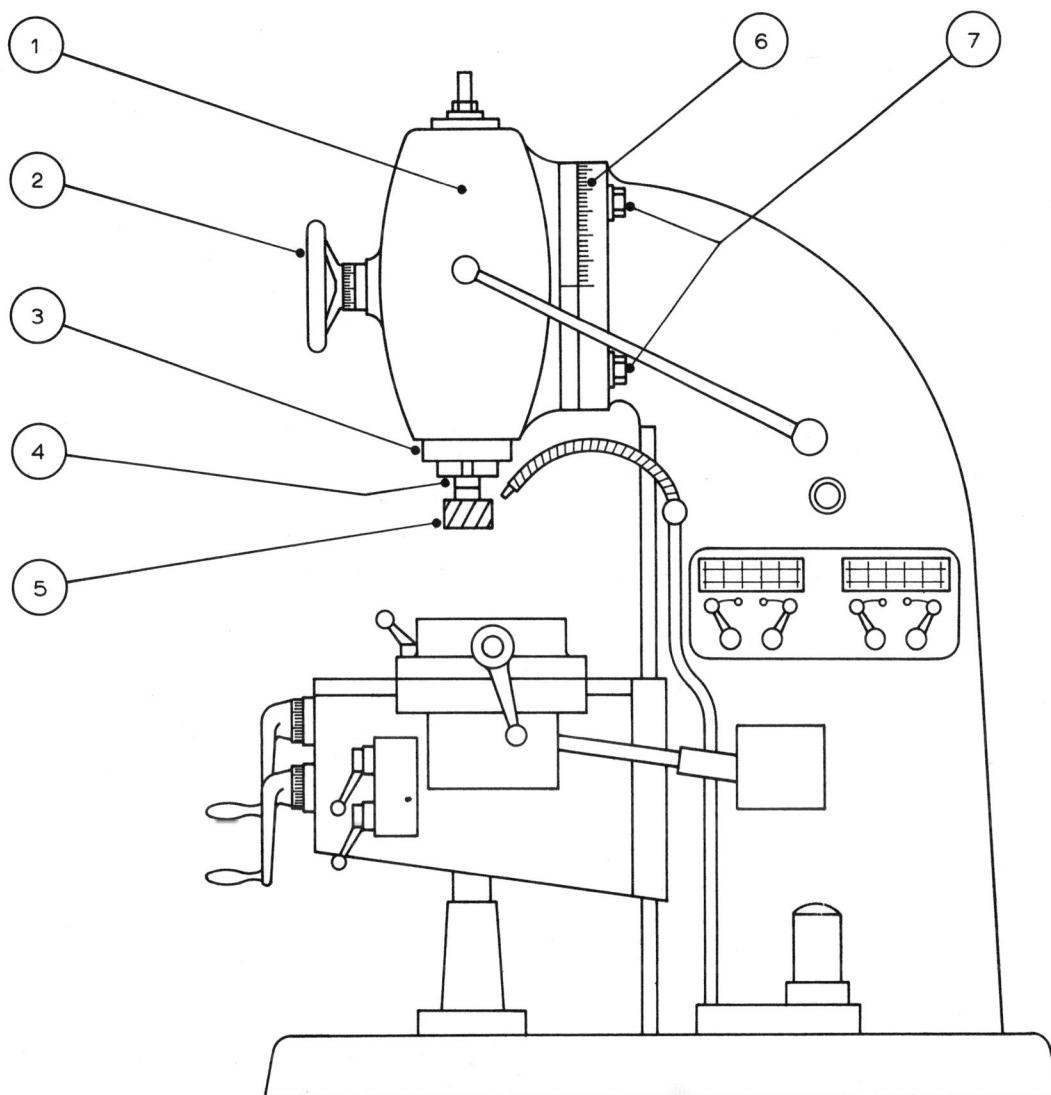

Parts of the Vertical Milling Machine

1 Vertical head which tilts.
2 Vertical feed handwheel.
3 Quill.
4 Spindle.

5 Cutter.
6 Head tilts here.
7 Head locking nuts.

Note: Remaining parts are similar to the horizontal-type milling machine.

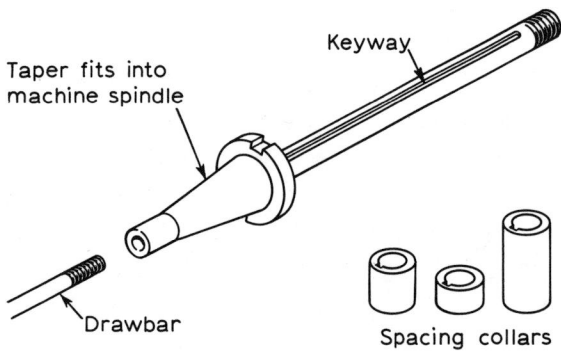

MILLING ARBOR

The Milling Machine

Holding the Cutter

The arbor shown on the left is as used on the horizontal milling machine. It consists of an accurately made shaft. One end has a taper which fits into a corresponding taper in the machine spindle. When the drawbar is tightened the arbor is held firmly in place.

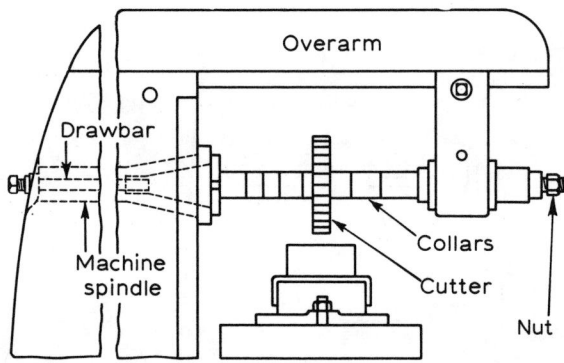

MOUNTING THE CUTTER ON THE ARBOR

The cutter is mounted in between spacing collars, as shown.
When the nut is tightened the collars hold the cutter in place. To prevent the cutter spinning on the arbor during use, a key is used.

Never

try to loosen the arbor nut without the support bracket in position—a bent arbor may result.

Always

clean the spacing collars before use.

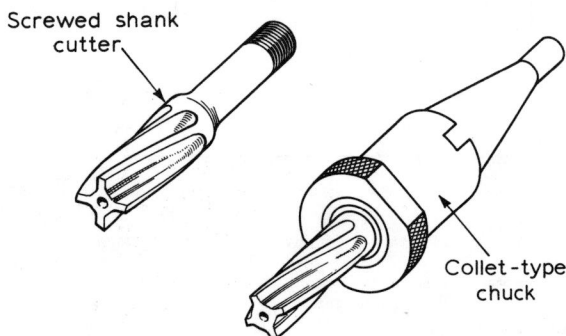

The collet-type chuck is used to hold cutters such as end mills, slot drills, etc. These cutters often have screwed shanks.
The chuck locates in the machine spindle in the same manner as the arbor, i.e. taper and drawbar. This type of chuck is used mainly on the vertical milling machine but can of course be held in the horizontal milling machine.

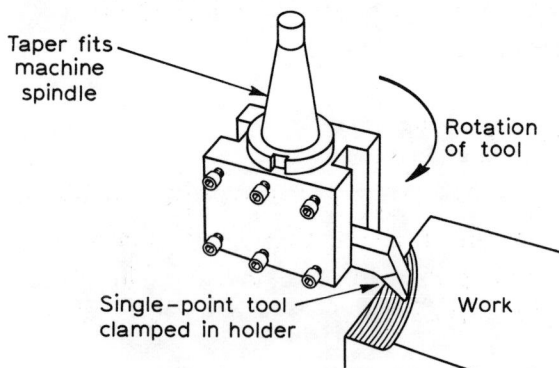

HOLDING SINGLE-POINT TOOLS

Holding of single-point tools is made possible by means of a special holder in which the tool can be clamped. This method of milling is suitable when large areas are to be cut. One advantage is that the tool is easily sharpened. This process is very often called 'flycutting'.

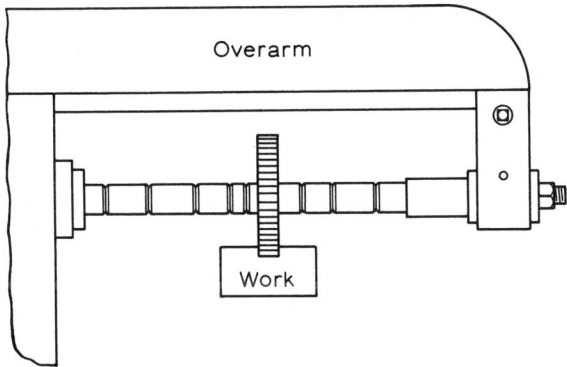

BAD

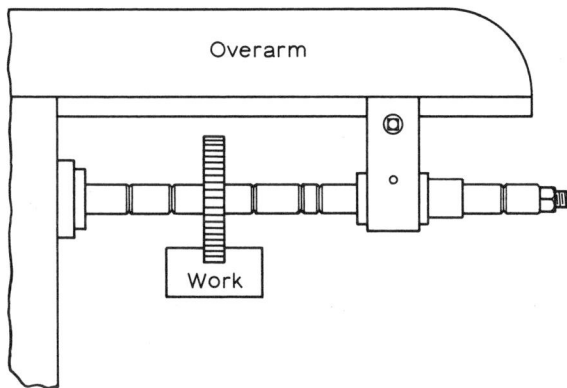

IMPROVED

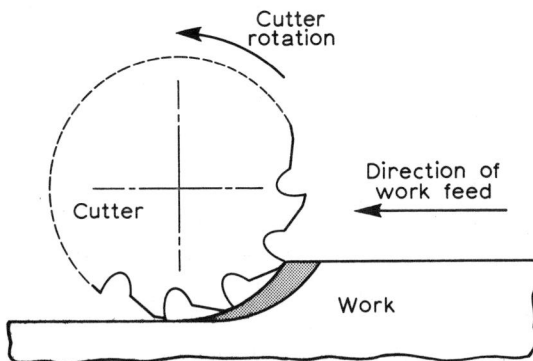

CONVENTIONAL OR UPCUT MILLING

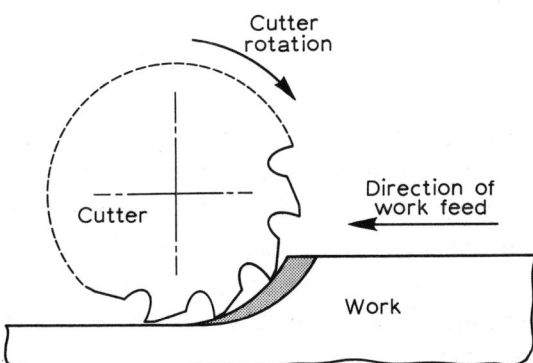

CLIMB MILLING

Rigidity of the Cutter

Rigidity is of great importance for efficient machining. The difference between a bad and a good set-up is shown on the left.
Always pay particular attention to the position of the arbor support. This should be as close to the column of the machine as possible to avoid deflection of the cutter.

Milling Methods

Conventional or Upcut Milling

This is the most common method of milling and is shown in the third illustration. Note the direction of rotation of the milling cutter. The cutting edge of the tooth can be seen cutting in an upwards direction. As the cutter is rotating in an anti-clockwise direction the work is fed in from the right.

Remember, the direction of rotation will determine which side of the cutter the work should be fed into. Conventional milling tends to lift the work away from the vice or machine table. Always ensure that the work is secure.

Climb Milling

Climb milling is shown in the bottom illustration. This method of milling should never be attempted unless under skilled supervision. There is always a tendency for the cutter to ride up or climb over the work. The advantage of this method is that the work is forced down on to the machine table when cutting takes place. Machines must be specially equipped to enable climb milling to be carried out safely.

Vertical Milling

Take care to feed the work at the correct side of the cutter. Remember, vertical cutters are unsupported at one end. Avoid climb milling on all but exceptionally light cuts.

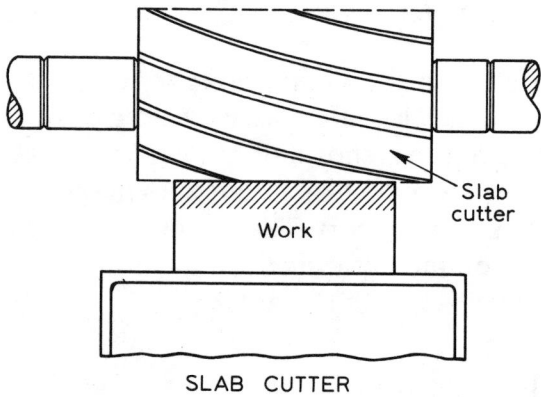

Milling Plain Surfaces

The Slab Cutter

This cutter is used on the horizontal milling machine for producing plain or horizontal surfaces. It is sometimes called a roller mill. Slab cutters vary both in diameter and width. Some have a large number of teeth and are used for light cuts. Others, with a small number of teeth and a sharp helix angle, are used for heavy cuts.

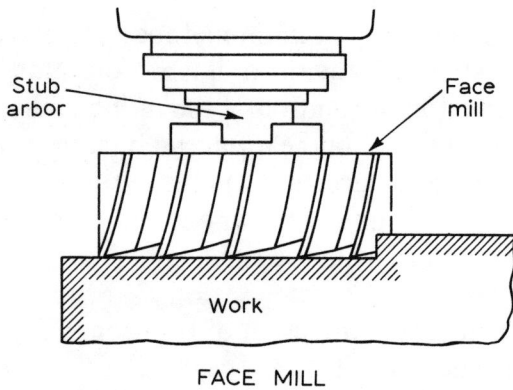

SLAB CUTTER

The Face Mill

This is also used for producing a flat surface. It can be used on the vertical or horizontal machine. A short arbor, known as a stub arbor, is used, on which is mounted the cutter. This type of cutter has teeth on the end as well as on the periphery.

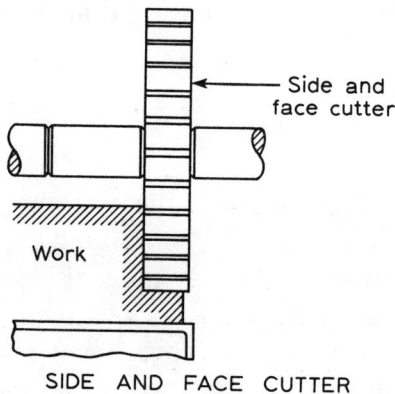

FACE MILL

The Side and Face Cutter

This cutter has teeth on the sides and also on the periphery. It is used to produce slots or for milling flats on round work held vertically.

If two cutters are mounted time can be saved, as two faces can be machined in one pass of the cutters.

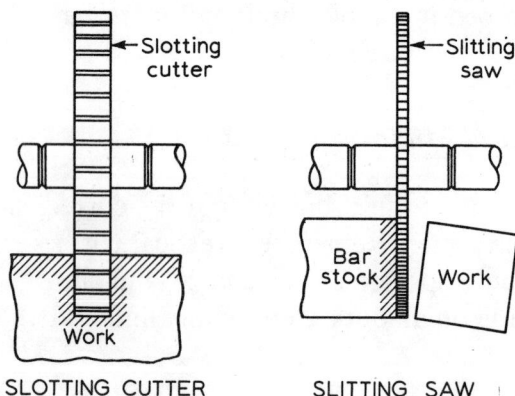

SIDE AND FACE CUTTER

The Slotting Cutter

This has teeth on the periphery only. It is used mainly for cutting slots or keyways.

The Slitting Saw

This saw is used for cutting material to length or for cutting narrow slots. Some cutters have teeth on the periphery only, in which case the cutter is thinner at the centre than at the edge. This will, of course, prevent binding of the cutter. Other types resemble very thin side and face cutters.

SLOTTING CUTTER   SLITTING SAW

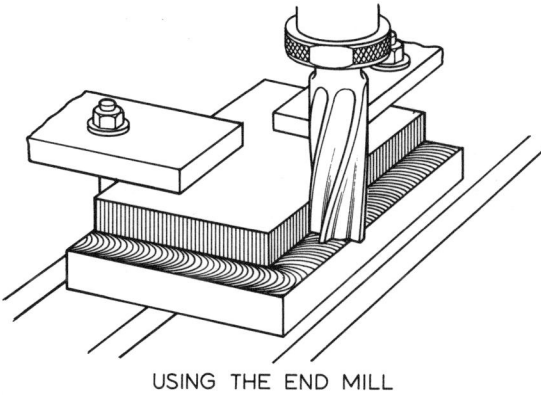
USING THE END MILL

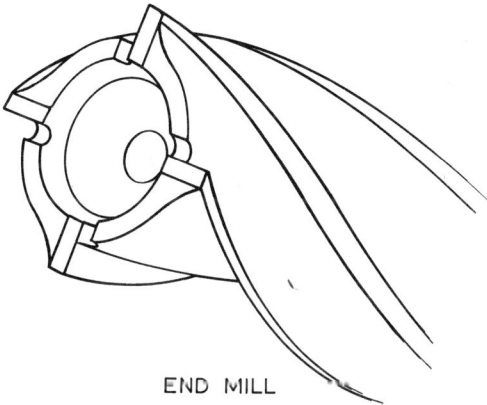
END MILL

The End Mill

The end mill is used mainly on the vertical milling machine to cut vertical or horizontal surfaces. An example can be seen in the top illustration.

End mills vary in size from approximately 3 mm diameter up to 40 mm diameter. End mills larger than this are supplied without a shank and are known as shell end mills. These are mounted on a stub arbor and are more rigid than conventional end mills. Due to the design of this type of cutter it cannot be plunged into the work in the same manner as a normal twist drill. The reason for this can be clearly seen in the second illustration. Cutting will not take place in the centre of the end of the cutter.

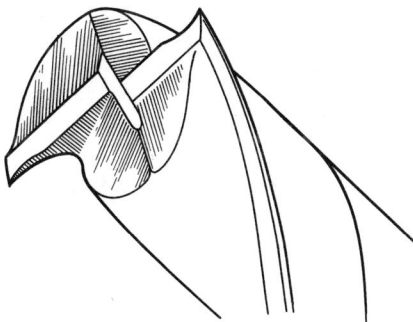
SLOT DRILL

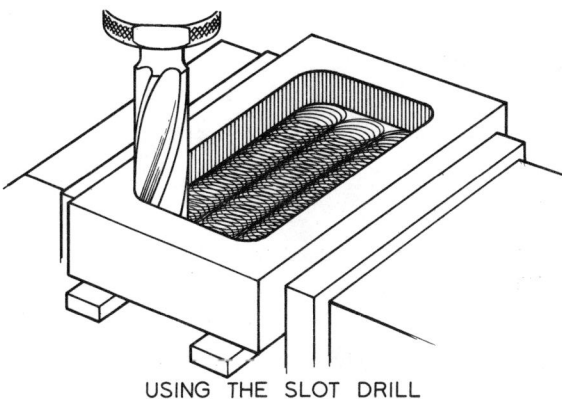
USING THE SLOT DRILL

The Slot Drill

The illustration on the left shows how the slot drill differs from the end mill. The cutting edges on the end of the cutter are made so that one edge is longer than the other. Cutting will therefore take place over the entire end of the cutter. This means that it can be fed into the work in the same manner as a twist drill. Cutting will, of course, also take place when using the side of the cutter. An example of the type of work this cutter is used for can be seen in the bottom illustration.

Reminder

Cutters only supported at one end will deflect if overloaded. Before you feed the work into an end mill or slot drill, check the rotation of the cutter and the direction of the feed.

Rate of Material Removal

Machine tools, such as lathes, shapers and milling machines, are expensive pieces of equipment and are of no value unless work is being produced. To produce a component, material must be removed. The removal of this unwanted material should always take place in the shortest possible time. The rate at which the material is removed will, of course, depend on the rigidity of the machine, the set-up and other factors. Large cuts will result in deflection of the work and therefore inaccuracy, as well as a poor finish, will occur. Material removal is normally divided into separate operations: (*a*) roughing cuts, and (*b*) finishing cuts.

Roughing

The operation of roughing is to remove as much material as possible in the shortest time. Common practice is to leave 0·5–1·0 mm for finishing cuts. Roughing cuts will normally take place at a higher rate of feed and greater depth of cut. It must be remembered that surface finish is of very little importance when roughing. Tool life is, of course, important and a large supply of the appropriate coolant is necessary. This is of particular importance when milling cutters are being used, because they are more difficult and take a great deal longer to sharpen than single-point tools.

Finishing

The amount of material to be removed is now small. Accuracy as well as surface finish may now be applied to the work. A word about surface finish: do not waste time producing a good finish or aiming at great accuracy if the drawing does not call for it. If the first cut produces a good finish but is found to be at the top end of the given tolerance it is not only unnecessary but a waste of time to take another cut in order to reach the nominal size (see page 69).

Summary

Material removal depends on:
- (*a*)   The material to be machined;
- (*b*)   The number of revolutions of the work;
- (*c*)   The rate of feed of the work or tool;
- (*d*)   The depth of cut.

Cutting Speed, Feed and Rake Angles

Turning

Listed below are the approximate cutting speeds for turning various materials using high-speed tools. The recommended rake angles are also shown.

Material to be Turned	Cutting Speed (m/min)	Side Rake	Back Rake
Aluminium	150–200	18–25°	15°
Brass	80–100	0–5°	0–3°
Cast iron	15–20	4–8°	4–6°
Copper	100–150	18–25°	12°
Hard steels	20–25	10–12°	6–8°
Mild steel	30–50	14–18°	8°
Plastics	50–200	*	*

* The rake angle for different types of plastic will vary considerably. To turn some plastics, little or no rake is required, whilst others are best machined with a large top rake. It is therefore recommended that advice is obtained before machining commences.

Reminder

The rate of feed applied to the turning tool will depend on the type of cut, i.e., roughing—coarse feed; finishing—fine feed.

Milling

Suggested cutting speeds using H.S.S. cutters.

Material to be Milled	Cutting Speed (m/min)
Aluminium	90–180
Brass	30–45
Cast iron	15–20
Copper	50
Hard steels	12–15
Mild steel	25–35
Plastics	30–150

Suggested feed per tooth for given cutters.

Side and Face Cutters (mm)	Slab Cutters (mm)	Slitting Saws (mm)	End Mills (mm)
0·25	0·40	0·12	0·25
0·20	0·25	0·10	0·25
0·25	0·30	0·12	0·20
0·20	0·25	0·12	0·25
0·12	0·15	0·10	0·12
0·15	0·25	0·10	0·12
0·25	0·40	0·12	0·25

The speed for a milling machine table is stated in millimetres per minute or inches per minute.

Note

Always keep the feed of a cutter as high as possible. This will not only lengthen the life of the cutter but will very often stop chatter and vibration. The same rule applies, of course, to turning and drilling operations.

B   Test Questions   B

Section B3

1   For what purpose are hand scrapers used?
2   Sketch three types of hand scrapers indicating clearly the cutting edges.

Section B4

3   List the two main types of lubricants giving three uses of each.
4   Sketch three types of lubrication systems giving examples of where each is used.
5   Why is cleanliness essential when lubricating machinery?
6   What health precautions should be observed when handling lubricants?

Section B7—Drilling

7   Sketch a twist drill indicating the following points:
    (a)   Body;
    (b)   Shank;
    (c)   Chisel edge angle;
    (d)   Point angle;
    (e)   Helix angle;
    (f)   Lip clearance angle;
    (g)   Lip length.
8   What is the effect of (a) unequal lip length and (b) unequal cutting edge angles on a twist drill? Use a sketch to illustrate your answer.
9   What is the purpose of the taper on the body of a twist drill?
10 Sketch two ways by which a twist drill can be held in a drilling machine.
11 Sketch three methods of holding a workpiece while drilling.
12 What safety precautions should be observed when drilling?
13 What is meant by each of the following terms? Use a sketch to illustrate each.
    (a)   Countersinking;
    (b)   Counterbore;
    (c)   Spotfacing;
    (d)   Reaming;
    (e)   Holecutting.
14 For what purpose are reamers used?

Section B7—Turning

15 What are main uses of the tailstock?
16 What is the purpose of:
    (a)   the feed shaft;
    (b)   the leadscrew;
    (c)   the compound slide;
    (d)   the apron;
    (e)   the chuck;
    (f)   the headstock;
    (g)   the tailstock.
17 What are gap beds and why are they used?
18 What is a four-way tool post and why is it used?
19 What is the essential difference between a three jaw and a four jaw chuck? Give examples of the types of job for which each is used.
20 Why are cutting fluids used?
21 Sketch:
    (a)   a parting-off tool;
    (b)   a boring tool; and
    (c)   a round-nosed roughing tool, indicating the rake and clearance angles.
22 Why is it necessary to correctly set the cutting tool in a lathe?
23 What is the effect on rake and clearance angle of (a) setting the tool high and (b) setting the tool low?
24 What is the effect of excessive overhang of a lathe cutting tool?
25 Why is a finer feed used on a finishing cut than on a roughing cut?
26 What will happen to the job if the jaws of the chuck are badly worn?
27 Why is it necessary to ensure that a lathe chuck holding a job is balanced.
28 Why are steadies necessary?
29 What is meant by each of the following terms? Use a sketch to illustrate each.
    (a)   Parallel turning;
    (b)   Facing;
    (c)   Boring;
    (d)   Parting off;
    (e)   Turning between centres.
30 Sketch a screwcutting tool.
31 For what is a hand chaser used? What safety precautions should be observed when using a hand chaser?

32 List the methods that can be used to turn a taper.

33 What safety precautions should be observed when operating a lathe?

Section B7—Shaping

34 Describe the cutting action of a shaping machine.

35 What is a clapper box and why is it used?

36 How is the workpiece held in a shaping machine?

37 What is a quick return mechanism and why is it used?

38 Sketch a shaping tool. Why does it have to be so robust?

39 Why is it necessary to have a table support on a shaping machine?

40 Compare a slotting machine with a shaper, stating the major differences.

Section B7—Milling

41 Sketch the cutter holding arrangements on (a) a horizontal milling machine and (b) a vertical milling machine.

42 Why should the unsupported arbor length be kept to a minimum on a horizontal milling machine?

43 How can a single-point tool be held in a milling machine?

44 What is the process called flycutting?

45 What are the basic differences between upcut and climb milling?

46 For what are each of the following milling cutters used:
    (a) Slab cutter;
    (b) Face mill;
    (c) Side and face cutter;
    (d) Slotting cutter;
    (e) Slitting saw;
    (f) End mill;
    (g) Slot drill.

47 Using sketches show how a slab cutter is mounted on an arbor.

48 What is a cutter adaptor and when is it used?

49 What is a stub arbor and when is it used?

50 What precautions should be taken when fitting an arbor into a milling machine?

51 Why should both the work and the cutter be set as near as is possible to the column of a horizontal milling machine.

52 Sketch the ways in which a workpiece can be secured to the table of a milling machine.

Section B7—Rate of Material Removal

53 What factors affect the rate of material removal of a machine tool?

54 A metal bar of diameter 120 mm is being turned at 35 rev/min. What is the cutting speed?

55 A bar of mild steel 50 mm diameter is to be turned at 40 metres per min. What spindle speed should be used?

56 The following information relates to a milling cutter:
    Diameter of cutter = 150 mm
    Number of teeth on cutter = 12
    Feed of cutter = 0·15 mm per tooth
    Cutting speed = 15 m/min
    Travel of cutter 200 mm
  Calculate the time per cut.

57 A milling cutter has 14 teeth and the feed per tooth is 0·2 mm. If the cutter is revolving at 100 rev/min, calculate the feed of the table in m/min.

C2   Health Hazards   C2

Precautions Against Toxic Substances

There are many substances used in industry which are dangerous to health. They may be in either solid, liquid or gaseous form. Strict precautions are observed to protect people, in factories or on sites, who come into daily contact with these health hazards. These precautions may take the form of:

1   Daily changes into freshly laundered clothing.
2   Provision of special ventilation equipment.
3   Use of respirators.
4   Use of protective clothing.
5   Periodic medical inspection.

Industries and their employees who are in daily contact with these substances are aware of the dangers involved and take the necessary precautions. There are, however, substances which are used in small quantities which can lead to a health hazard over a period of time. These substances should be known to the worker in order that he can take the necessary safety precautions. The following irritate the skin and can lead to dermatitis or, at the worst, to malignant skin complaints:

Mineral oils.
Cutting fluids.
Solvents (paraffin, trichlorethylene, turpentine, petrol, etc.).
Coal tar products (cresols, phenols, etc.).
Alkalis.
Acids.
Chromates.

Bichromates.
Naphthalines.
Diphenols.
Certain insecticides.
Resins.
Powders (cement, flour, etc.).
Certain hard woods.
x-rays, beta- and gamma-rays.
Intense heat and light.

These act in different ways: an acid may severely burn the skin, producing ulcers which can be slow to heal; solvents dissolve the natural oils in the skin, making it susceptible to cracking with possible bacterial infection; radiation destroys the cell tissue. The obvious safety precautions to give complete skin protection against them are:

1   Use of gloves of various materials, depending upon the substance to be handled.
2   Use of face masks or possible respirators.
3   Eye protection, which is essential against corrosive substances or harmful irritating fumes.

Certain substances are extremely dangerous when inhaled. For example, when preparing lead pipes for soldering, the pipes must never be cleaned with abrasive material such as sanding discs or emery cloth, as the fine lead dust produced can cause pulmonary troubles, or be absorbed into the digestive tract resulting in poisoning. Lead should always be cleaned by scraping with a knife.

Workers should at all times be aware of the importance of cleanliness. Good washing facilities are essential and should be used before taking a meal. It is not unknown for a workman to be poisoned by eating sandwiches with unclean hands which have become contaminated with such materials as lead or arsenical compounds. Fumes from pickling, soldering, and certain welding operations can be irritating and precautions should be taken against them.

C6   Heat   C6

Measurement of Temperatures in the Workshop

Many of the processes carried out in the workshop, e.g. annealing, normalizing, pre-heating, post-heating, etc., require components or materials to be heated to certain temperatures. The methods used to measure these temperatures may have to be precise or an approximate value may be permitted, dependent upon the circumstances. The following are methods of measuring temperatures approximately:

Colour	Approximate Temperature (°C)
Brilliant white	1 500
Bright white	1 400
White	1 300
Bright orange	1 200
Orange	1 100
Bright cherry red	1 000
Cherry red	900
Brilliant red	800
Dull red	700
	600
Faint red	500

Colour Changes of Metals

Some metals, such as steel, have noticeable colour changes as their temperature alters, each colour indicating an approximate temperature range (see chart on the left).

In the case of copper the only noticeable effect is the change in the degree of redness, i.e. from a dull red to a light orange. Other metals, such as aluminium, have no noticeable colour change and this method cannot be applied to them.

Heat-sensitive Crayons or Paints

Heat-sensitive materials in the form of crayons or paints are available to cover temperature ranges from 45°C to 1 370°C.
When metal objects suitably marked with one of the above materials attain the appropriate temperature, the marking material changes its colour, e.g. a brown mark will change to green at 350°C. Once this change in colour has taken place the heat-sensitive material will not change colour again regardless of any temperature change.

Segar Cones

These cones are small pyramids of material, the composition of which depends on the temperature to be indicated. They are placed upright in the furnace and when the desired temperature is reached the top of the pyramid arches over until it is almost level with its base. A further increase in temperature will cause the material to collapse and form a shapeless mass, as illustrated.

It is usual to use three cones, one at the desired temperature, the other two being approximately 20°C above and below this temperature. It is important when using segar cones to ensure that the furnace is not heated too rapidly.

The following are methods of precise temperature measurement:

Liquid Expansion Thermometer

The most common types of liquid expansion thermometer are the mercury-in-glass, for measuring up to 500°C, and the mercury-in-steel, for measuring up to 600°C. The mercury-in-steel type is used in conjunction with a fine flexible capillary tube connected to a Bourdon-type pressure gauge (shown on left), calibrated to read temperature. This gauge can be fitted up to 30 m from the thermometer bulb.

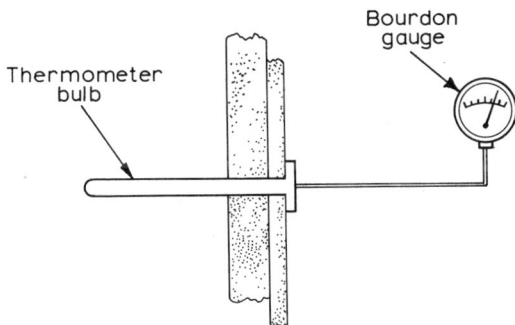

Vapour Pressure Thermometer

These thermometers are similar to the mercury-in-steel type but use a volatile liquid. They can be used to measure temperatures up to 800°C and the gauge can be situated up to 60 m from the thermometer bulb.

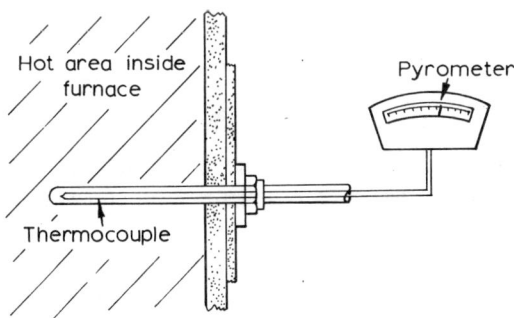

Thermocouple Pyrometer

If two different metals are joined together and heat is applied at the joint, an e.m.f. is produced, causing a current to flow. A meter connected into this electrical circuit will measure this e.m.f., the value of which is proportional to the temperature of the joint. The meter is calibrated to read temperature.

If this joint or thermocouple as it is called is placed in contact with, or in the vicinity of, heat (e.g. within a furnace), the temperature of the component can be measured accurately.

Optical Pyrometer

This type of pyrometer compares the intensity of light radiated from the heat source with a known source.

A typical type is the disappearing-filament pyrometer which employs an electric lamp filament, the light intensity of which can be varied by a rheostat until it matches the heat source to be measured. The temperature is indicated on an ammeter calibrated to read temperature over a range of 700–1 800°C.

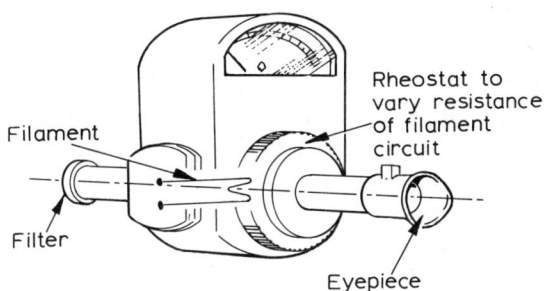

Temperature Measurement for Tempering, Hardening, Forging, Soldering, Brazing and Welding

Tempering Temperatures for Steels

Tempering Colour	Temperature (°C)
Pale straw	230
Dark straw	240
Brown	250
Brownish purple	260
Purple	270
Dark purple	280
Blue	300

Tempering

When being tempered, metal should be heated to between 180°C and 650°C, depending on the properties required. To achieve this it is usual to place the metal in a temperature-controlled furnace, oil baths, salt baths or lead baths so that the temperature can be controlled accurately. In the case of tools which are normally tempered at low temperatures, it is possible to use the colour of the surface to estimate the temperature, but it is important to remember that this is only an assessment of the temperature.

Hardening

As with tempering, when hardening it is necessary to heat metal to an exact temperature, depending on the desired properties and the type of material being dealt with. It is therefore usual to heat the metal either in a temperature-controlled furnace or a furnace whereby the temperature can be estimated by some visual means, such as an optical pyrometer.

Forging

Forging is carried out on steels which have been heated to a bright red heat, the forging process being continued until the metal has cooled to a dull red heat. If necessary, temperature-controlled furnaces or pyrometers can be used to give more accurate temperature readings for carrying out this operation.

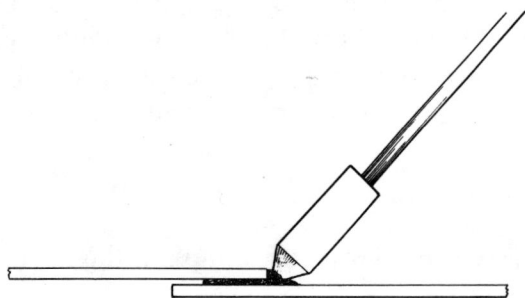

Soft Soldering

The metal, when soldering, must be brought to a sufficiently high temperature to allow the solder to run freely over the metal and to achieve maximum adhesion and penetration, as shown on left. With a little practice the operator will have no difficulty in estimating the correct soldering temperature. These temperatures vary in accordance with the type of solder being used, but are usually within the range 180–250°C.

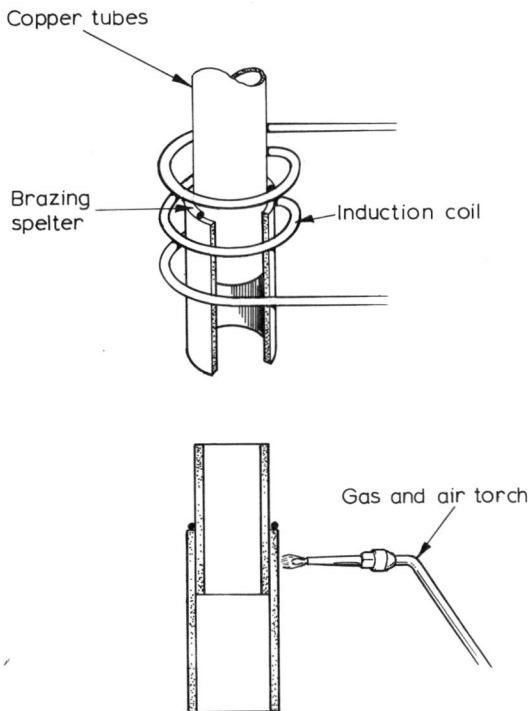

Copper tubes

Brazing spelter

Induction coil

Gas and air torch

Weld metal

Good fusion

Good penetration

Brazing

This process is carried out at a higher temperature range, i.e. 630–1 083°C. The importance of the correct temperature is the same as for soldering, i.e. it allows fluidity of brazing alloy, good adhesion and penetration. Various heating sources (see the top two illustrations) are used for brazing, from temperature-controlled induction coils to gas and air torches. In the latter case an experienced operator will have no difficulty in assessing the correct temperature.

Welding

During welding, the metal must be brought to a temperature high enough to melt both parent and filler metal, to allow complete fusion between the two, as shown in the bottom illustration.

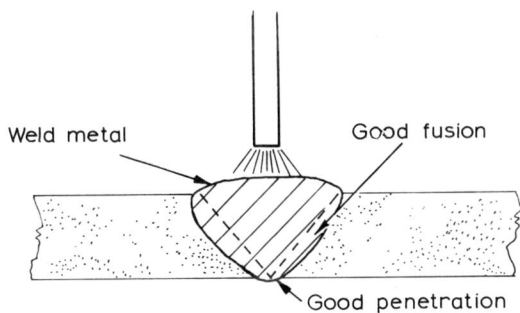

If the process is a manual one, e.g. metal arc or oxy-acetylene welding, the operator would see the metal becoming molten and know when the correct temperature is being reached. On an automatic process such as submerged arc, however, the machine must be set to the manufacturer's recommended instructions to obtain the temperatures necessary to produce a sound weld.

The Practical Effects of Expansion and Contraction of Metals

When metals are heated they expand and increase in length. Similarly, when cooled they contract. The amount by which they expand or contract depends on three factors:

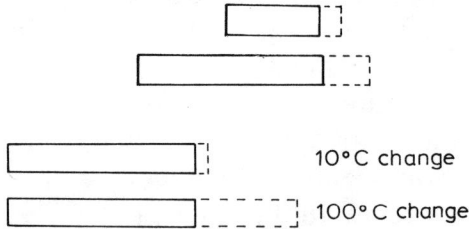

10°C change

100°C change

Value of Coefficient of Linear Expansion

Metal	Value (°C)
Aluminium	0·000 023 8
Brass	0·000 020 0
Copper	0·000 016 7
Steel	0·000 011 2
Cast iron	0·000 010 6

1 How long they are to start with—as you would expect, a bar twice as long as another will expand twice as much.

2 The amount the temperature changes—bigger changes in temperature will give greater changes in length than smaller temperature changes.

3 The type of metal—different metals expand at different rates. A constant value for each material is available. It is called the Coefficient of Linear Expansion (see table on left).
Change of length = Original length × Coefficient of Linear Expansion × Temperature change.

Example

If a steel plug, gauge 50·00 mm diameter, has its temperature raised from 20°C to 27°C, what will be its new diameter at this temperature?

Coefficient of Linear Expansion
  (from table)   = 0·000 011 2
Change of temperature = 27−20 = 7°C
Change in diameter   = 50·00 ×
                         0·000 011 2 × 7
                  = 0·003 9 mm
New diameter at 27°C = 50·00+0·003 9
                  = 50·003 9 mm.

This property of metal to expand and contract can be used to advantage or disadvantage, as follows:

Used to Advantage

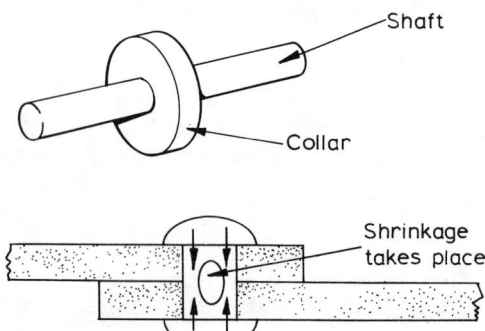

Shaft

Collar

Shrinkage takes place

1 It can be used to shrink components together, such as a shaft or collar which has an interference fit. The collar is heated, it expands and slips on to the shaft; when it cools it shrinks and fits tightly on the shaft.

2 After heat-riveting has taken place the centre of the rivet is still hot and, on cooling, tends to shrink and pulls the plates together.

Circuit broken due to expansion

Copper strip

Iron strip

LENGTH AND SHAPE BEFORE WELDING

LENGTH AND SHAPE AFTER WELDING

Weld unable to contract
due to heavy sections
restricting movement

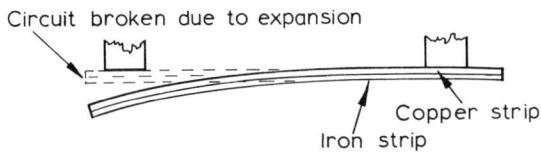

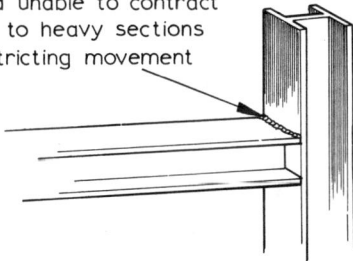

3 A bimetal strip is a strip of two different
   metals joined together. When it is heated
   it curves, because the two materials
   expand at different rates. It is used in
   thermostats to open and close an electric
   circuit.

Used to Disadvantage

1 It can cause distortion of welded
   components and create a change in their
   dimensions due to shrinkage.
2 It can cause internal stress to be set up,
   resulting in cracking of the component.
3 Measuring instruments are calibrated
   to be used at a particular temperature.
   If they are used at other temperatures they
   will not measure accurately.

Concept of Heat Capacity

Heat is a form of energy, and the units in which it is measured are fundamental units of energy.
There is no real reason why we should not use the unit *joules* to measure quantities of heat energy,
but it is more convenient to use specially derived units. These units are:
  (*a*) the *calorie* (cal), which is the amount of heat energy required to raise the temperature of
         1 gramme of water by 1°C;
  (*b*) the *kilocalorie* (kcal), which is the amount of heat energy required to raise the temperature
         of 1 kilogramme of water by 1°C. The kilocalorie is equal to 1 000 calories.

Quantity of Heat

The amount of heat energy required to change the temperature of a piece of material depends upon three factors:

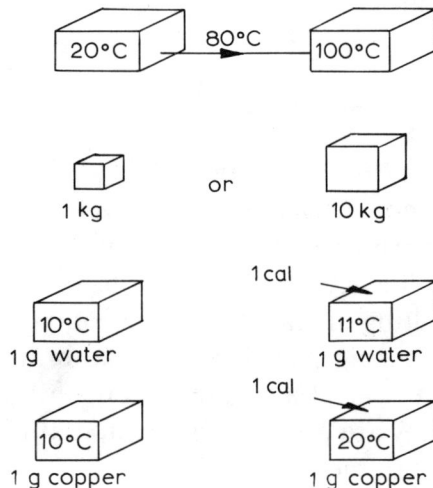

1   The required change in temperature of the body.

2   The quantity of material in the body, i.e. its mass.

3   The type of material the body is made of. (If we supply 1 cal of heat to 1 g of water its temperature will rise by 1°C, but if we put this 1 cal of heat into 1 g of copper its temperature will rise by approximately 10°C.)

Specific Heat of Various Materials

Material	Specific Heat
Aluminium	0·211
Copper	0·093
Lead	0·031
Steel	0·117
Solder	0·043
Oil	approx. 0·47

To allow for this difference, which will occur with all the different materials that we have to deal with, we have a set of constants which compare this heat capacity of the different materials with that of water. This constant is called the *specific heat* of the material. The table opposite gives the specific heat values for some of the more common materials.

The *quantity of heat* is calculated by:
Quantity = Mass of material × Specific heat of material × Temperature change of material

Example

Calculate the quantity of heat in calories required to heat 5 kg of copper from 20°C to 420°C.

Required change in temperature = 420−20
                              = 400°C
Specific heat from table   = 0·093
Quantity of heat = 5×1 000×0·093×400
                 = 186 000 cal.

If we were required to give the answer to the above question in joules we would have to use the conversion factor 4·18 j = 1 cal. Therefore 186 000×4·18 = 777 480 j.

C7   Change of Physical State of Materials due to Temperature Changes   C7

Hardening and Annealing Temperatures

Carbon Content of Steel (%)	Temperature (°C)
less than 0·1	875–925
0·12–0·25	840–870
0·25–0·5	815–840
0·5–0·9	780–810
0·9–1·3	760–780

HARDENED

HARDENED AND TEMPERED

Quench Hardening and Tempering of Steels

When steel is being hardened it is heated to its annealing temperature (see table on left) and then rapidly cooled by quenching. The quenching medium is usually oil or water, but if a very rapid quench is required a solution of salt or caustic soda and water can be used. It is important to remember, however, that if the cooling rate is too drastic, cracking or distortion of the parts being hardened will result. As can be seen in the table, the temperature to which the steel must be heated depends upon the carbon content of the steel, as also does the degree of hardness achieved. It is not usual to harden steels by this method if they have less than 0·3% carbon content.

Tempering

The hardening process discussed above produces steels which are extremely hard and have good wear-resistant properties, but they are also very brittle. To reduce this brittleness the material has to be tempered (it is useless having a very hard cutting tool that will splinter on its first use).

The tempering process requires the metal to be reheated and then either quenched or allowed to cool naturally. Tempering removes some of the hardness and also increases the toughness. The temperatures for tempering are shown in the table below.

Tempering Temperatures

Temperature (°C)	Toughness (increased)	Hardness (decreased)	Application
220			Turning and shaping tools
240			Milling cutters and drills
250			Taps
260			Reamers, twist drills, punches and dies
280			Cold chisels
300			Wood saws, springs
450–600			Structural steels

Some tools, such as cold chisels, only require the cutting edge to be hardened. To do this, the tool is heated to the required hardening temperature and then the cutting end only is quenched. When cool, it is then reheated to the required tempering temperature and quenched. This ensures that the chisel has a hard cutting edge and a tough body.

Practical Effect of the Variations in the Carbon Content of Steels in the Normalizing, Annealing, Hardening and Tempering Operations

In the heat-treatment processes the percentage of carbon present in a steel is one of the important factors in determining the required process temperature and the resulting properties of the steel.

Normalizing

In normalizing, the carbon content not only determines the temperature to which the metal must be heated (i.e. temperatures similar to those used for annealing), but also produces a steel which is slightly harder and stronger than a fully annealed steel of the same condition.

Annealing

Annealing is carried out on high-carbon steels to refine their structure prior to hardening, as otherwise these steels would crack during quenching.

In general, the higher the carbon content of the steel the more severe are the internal stresses it contains as a result of its previous treatment by rolling, forging, etc., and therefore the heating rate should be slower.

Hardening

For hardening, the temperature to which the steel should be heated varies according to the percentage of carbon present in the steel.

From pure iron which contains no carbon to steel containing 0·87% carbon this temperature decreases relatively slowly from approximately 950°C to 750°C.

The temperature rises steeply with an increasing carbon content above 0·87%, with an increased danger of brittleness and cracking.

Steels containing less than 0·3% carbon cannot be hardened effectively, the maximum effect being obtained in steels having about 0·7% carbon content.

Tempering

Tempering is usually only carried out on high-carbon steels which have been subjected to a hardening process.

Low-carbon steels are not easily hardened and therefore tempering is not required.

Change of Physical Properties due to Forging Processes, Welding and Cutting

Many carbon steel forgings are made from bars and billets up to and including 100 mm square or diameter, which necessitates heating as a preliminary to the forging process. If the finishing temperature of the forging is correct then grain refinement is obtained with an improvement in strength, ductility and toughness, but directional properties are produced. The improvement is greater in the direction of working than in the transverse direction (see top illustration).

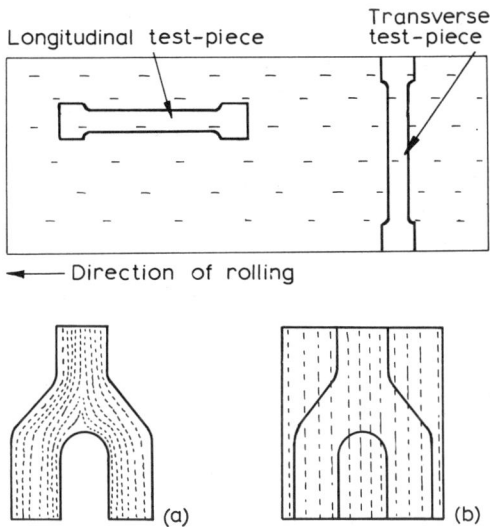

In forging, the 'fibres' should follow the shape of the section (see (a), second illustration).

Forgings are, therefore, superior in mechanical properties to similar shapes which have been machined from hot-rolled material (see (b), second illustration).

When a steel is cut by an oxy-fuel gas process, a large quantity of heat is liberated in the cut thereby heating the metal adjacent to the cut to a very high temperature. As the heat moves forward fairly quickly the mass of cold metal acts as a quenching medium and rapidly cools the heated metal. The steel will therefore harden. The degree of hardness depends on the amount of carbon and alloying elements present and the thickness of the material being cut (see table on left).

Thickness (mm)	Approx. Depth of Hardness (mm)	
	Lower carbon Steels	High-carbon Steels
<12	<0·7	0·7
12	0·7	0·7–1·4
150	3·0	3·0–6·0

The hardness effect decreases as the distance from the cut increases. For low-carbon steels (i.e. 0·25% carbon and under) the cut surface is at least as equal in physical properties as the plate itself.

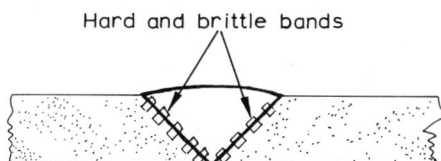

The change in physical properties that takes place when steel is welded depends on:
(a) the composition of steel, and (b) the welding process used.

Low-carbon steels may be welded with very little difference in strength, ductility and hardnesss between the parent plate and the weld metal. Medium- and high-grade steels, however, may result in a hard and brittle band forming along the line of fusion between the weld and parent metal (bottom illustration). This is due to rapid cooling when welding by an electric-arc process, without taking the necessary precautions. There is also a marked decrease in ductility and elasticity under these conditions.

Multi-run deposits on thick sections tend to normalize each preceding deposit to increase the strength, ductility and hardness of the weld metal.

Effects on Copper and Aluminium Alloys of (a) Forging Processes, (b) Welding, and (c) Cutting

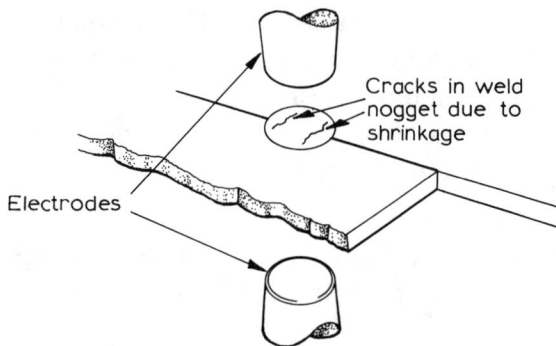

Cracks in weld nogget due to shrinkage

Electrodes

Forging, and processes such as resistance welding, can be carried out on aluminium without any deterioration in the strength of the metal. Aluminium shrinks considerably on cooling and this can lead to cracking (top illustration). Overheating can cause over-ageing and annealing.

With copper, which possesses high conductivity, it is difficult to bring the fusion faces only to the correct temperature, and overheating will result in annealing and a decrease in ductility of the metal.

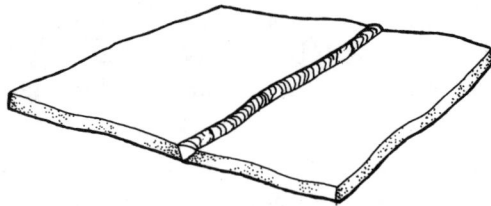

Welding aluminium could result in excessive distortion unless adequate precautions are taken (second illustration).

Certain of the aluminium alloys, e.g. duralium or alclad, may also suffer a decrease in strength with an increased risk of cracking. Also, the corrosion resistance may also be lowered, as a result, to the welding heat.

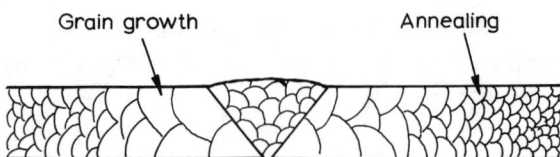

Grain growth   Annealing

The welding of copper may result in annealing or grain growth taking place in the weld zone (third illustration), with an accompanying lowering of its mechanical properties.

Flame-cutting copper and copper alloys requires extensive preheating due to copper's high conductivity, resulting in a softening of work-hardened metal and danger of grain growth in metal which is in the annealed state.

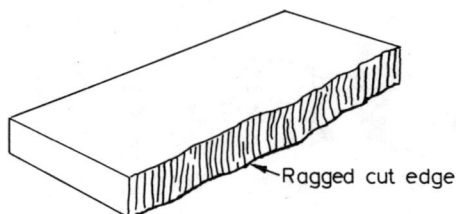

Ragged cut edge

On aluminium, the quality of the cut is poor, leaving a ragged profile (bottom illustration). Certain of the aluminium alloys will develop a hard surface due to the formation of oxides during the flame-cutting process.

C8   Non-Metallic Materials   C8

By far the most common materials used in engineering are based on iron, for example, high-, medium- and low-carbon steels. The majority of engineers, whether designers or shop floor workers, are concerned with this group of alloys. However, this is not to say that other materials are not of equal importance because they are not used so frequently. Other metallic materials, such as aluminium, copper, tin, lead, zinc, nickel, chromium, titanium, and the alloys formed from them, are extremely important to all branches of engineering.

All materials have certain advantages and disadvantages. The metallic materials all suffer from various forms of corrosion and it is in this field, in particular, where non-metallic materials, particularly plastics, have proved their importance.

Plastics are the product of chemical synthesis and at some stage of their production are capable of flow. Under the application of heat and pressure they can be caused to take up a desired shape, which will be retained after the forces have been withdrawn. Plastics fall into two main categories:

1   Those whose structure preserves their independence and can be softened and resoftened by heat—these are called *thermoplastic*.
2   Those whose structure becomes rigid by 'cross-linking'—these are called *thermosetting*.

Thermoplastics

1   *Cellulose nitrate:* celluloid—films and lacquers, explosives.

2   *Cellulose acetate:* non-inflammable celluloid—injection moulding.

3   *Ethenoid plastics:*
   (*a*)   Polytetrafluoroethylene (PTFE)—gland packings, pipe jointings, valve seatings; corrosive resistant.
   (*b*)   Polythene, polyethylene—fabrication and lining of chemical equipment, vessels, etc., acid-resisting floors, dishes, buckets, etc. By varying the basic ethylene structure we obtain:
   (*c*)   Vinyl chloride (PVC)—cable protection, switch boxes, junctions, television and radio components, pipes, sheets. By substituting chlorine we obtain:
   (*d*)   Styrene (in an expanded form, air is trapped giving heat and sound insulation)—panels, sheets, tiles. By substituting phenol we obtain:
   (*e*)   Perspex—transparent and translucent panels, windows, etc. By substituting methyls and carboxymethyl we obtain:
   (*f*)   Nylons (super polyamides)—fibres, bearings, bushes; high abrasion resistance and ability to run at high speeds under light loads without lubrication.

Thermosetting

1   *Phenolic plastics:* expanded high-pressure laminates—building, structural work, aircraft, ships, etc., sheets, rods, tubes; fabric reinforced, gear wheels, water-cooled bearings.

2   *Amino plastics*—moulding.

3   *Polyester plastics:* quick-curing compression-moulding powders, low-pressure laminates.

4   *Allyl plastics:* contact laminates, glass fibre.

5   *Silicon-based plastics:*
   (*a*)   Silicon fluids—stable against heat oxidation and mechanical shearing; good lubricant between common bearing metals, particularly at high and low temperatures; good electrical properties.
   (*b*)   Silicon resins—paints and enamels to withstand continuous temperatures up to 260°C.
   (*c*)   Silicon rubber—water repellent, will withstand temperatures up to 150°C; used for electrical connections.

Working Plastics

Machining

Both thermosetting and thermoplastic types can be machined without difficulty using special tools on standard machines. There are differences in technique between plastics. A general guide is to treat them as if they were brass or copper.

Welding and Fabrication

Thermoplastic sheets can be fabricated and welded similar to light metals. In the main, welding is effected by fusion of the sheets by direct heating with hot air or gases, or by hot pressing between pressure surfaces. Sheet steel is now widely used, produced with a plastic coating. This steel can be pressed and formed without injury to the plastic facing.

Advantages and Problems Arising from the Use of Plastics instead of Metal for Pipes and Conduits

In modern industry the uses of pipes and conduits are numerous for the conveying of:

Liquids—water, acids, liquid chemicals.
Gases—oxygen, town gas, methane, etc., and ventilation fume extraction.

In the above cases the material being conveyed may be either hot or cold or under pressure.
Electrical cables can be grouped together and protected by conduits.
Plastic pipes, tubes and conduits can be used in certain cases because of the advantages plastics have when compared with metals.

The majority of pipes are made from steel, which is relatively cheap, but they must be protected from atmospheric corrosion by painting, galvanizing or coating with plastics. If they are painted it must be remembered that they will require regular maintenance and inspection.

When corrosive liquids are to be conveyed, more expensive metals will be required such as stainless steel, copper, nickel, etc.

Plastic piping offers a cheaper method of conveying corrosive liquid, it does not require painting, etc., and being light in construction will require lighter and cheaper supports. It must be remembered, though, that plastics will not withstand high temperatures and pressures demanded in some applications.

In the case of conduits made from plastics, they are exceptionally light and easy to erect, and give protection to the cables from corrosive conditions. Their main drawback is that the conduit cannot be used as an earth conductor, which means that an earth conductor must be introduced into the conduit as an additional conductor.

Advantages and Disadvantages of using Plastic Pipes and Conduits compared with Metals

Possible Advantages	*Disadvantages*
Light and easy to handle.	Unsuitable for high pressure.
Speedily erected.	Unsuitable for high temperature.
Does not corrode in air and requires no surface protection.	Unsuitable for conducting electricity.
In general, resists corrosive chemicals and is unaffected by frost.	

C   Test Questions   C

Section C2

1   Why should solvents never be used to clean the skin?
2   What protective clothing should be worn when handling the following substances?
    (*a*)   Mineral oil.
    (*b*)   Solvents.
    (*c*)   Insecticides.
    (*d*)   Resins.
    (*e*)   Acids.
    (*f*)   Cement powder.
3   Why should dry abrasive material never be used on lead?

Section C6

4   What are heat-sensitive crayons and how are they used?
5   What are segar cones used for, and how are they used?
6   List three methods of precisely measuring temperature in the workshop.
7   What determines the temperature required when soldering a joint?
8   During manufacture, the temperature of a brass component rises by 10°C above room temperature. At this temperature it measures 150 mm.
What will be its length when it cools back to room temperature?
9   A steel girder 20 m long has its temperature increased by 25°C. What will be its new length?

10   List three advantages and three disadvantages of the expansion and contraction of metals due to temperature changes.
11   Complete the table at the foot of the page.

Section C7

12   What is the effect of heating 0·7% carbon steel to 800°C and then quenching it in water?
13   The procedure above results in a brittle material, explain how this can be reduced.
14   Describe the heat-treatment process that would be given to a cold chisel, stating the reasons for the various processes.
15   Why are forgings superior in mechanical strength to similarly shaped components machined from hot-rolled material?
16   What is the effect on medium- and high-carbon steels of flame-cutting and welding?

Section C8

17   What is meant by a thermoplastic? Give five examples and uses of thermoplastic materials.
18   What is meant by a thermosetting plastic? Give five examples and uses of this group of materials.

Material	Mass	Temperature Change (°C)	Specific Heat	No. of Heat Units
Steel	10 kg	1 000		kcal
Copper	1·5 kg			930 kcal
Lead		300		1 860 kcal
Brass	6 kg	900		kcal
Oil	250 g	80		cal
Water	30 kg	70		cal
Aluminium	100 kg			25 320 kcal

D2   Engineering Drawings   D2

Interpretation of Engineering Drawings in Third Angle Projection

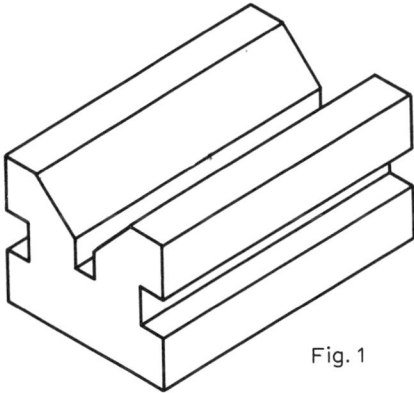

Fig. 1

Fig. 1 shows a pictorial view of a vee block, and Fig. 2 shows two views of the same block drawn in *third angle* projection. Fig. 3 shows three views of the dovetail block shown at Fig. 4. In this case, three views are necessary to show all the detail.

From the pictorial views shown on the following three pages, draw the necessary views in *third angle* projection.

Face A

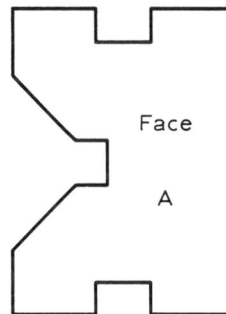

Face A

Fig. 2

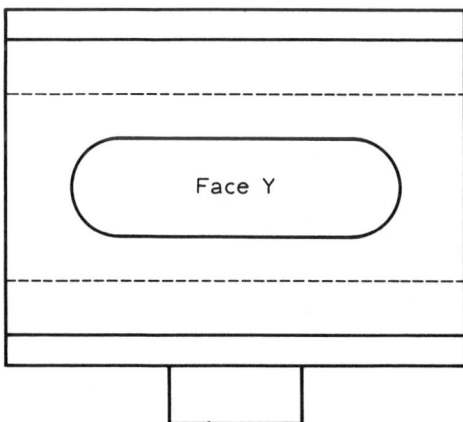

Face Y

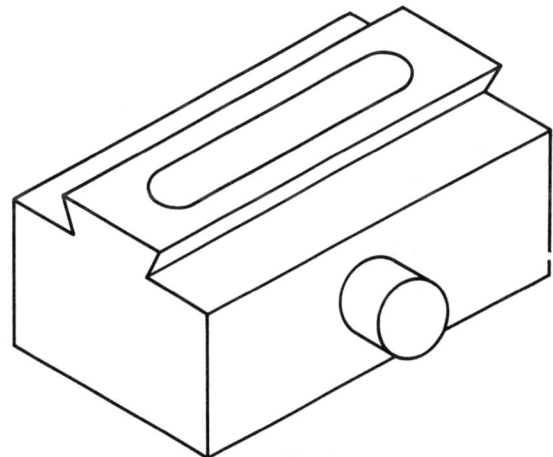

Fig. 4

Fig. 3

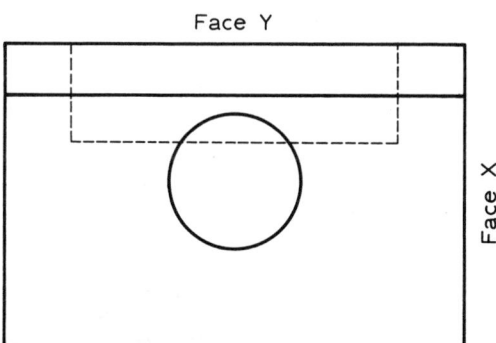

Face Y

Face X

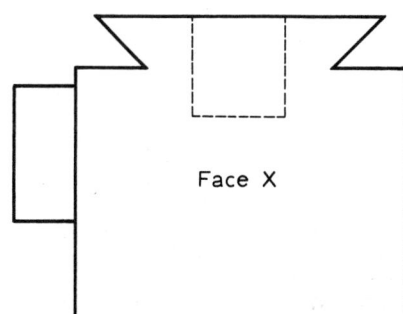

Face X

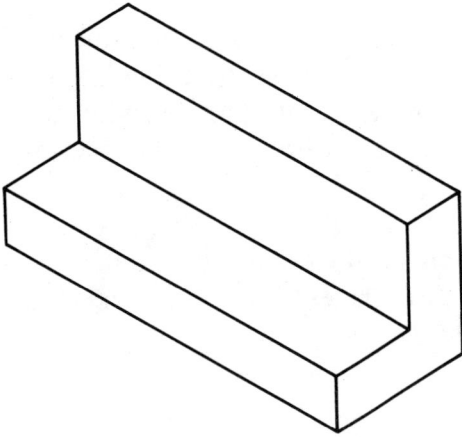

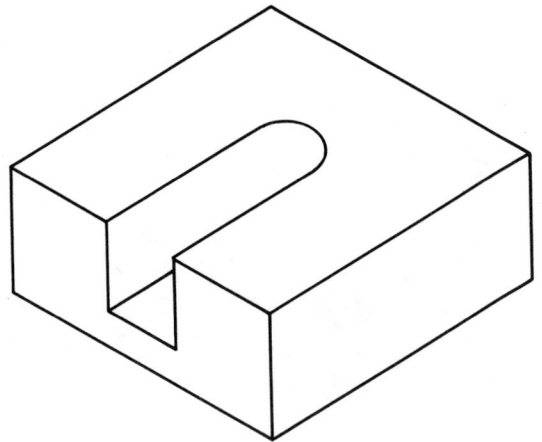

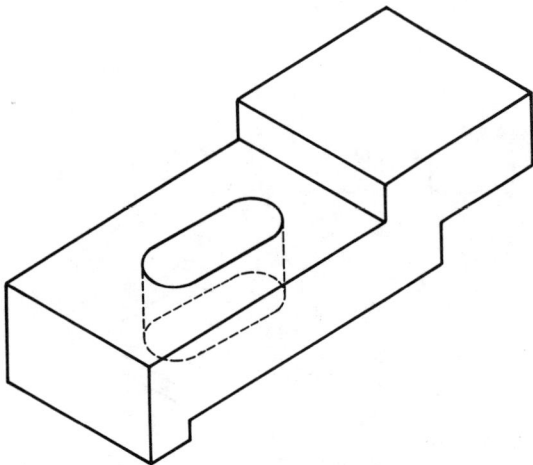

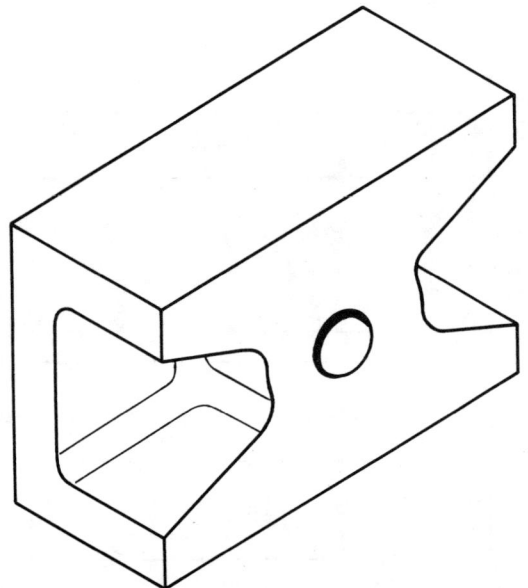

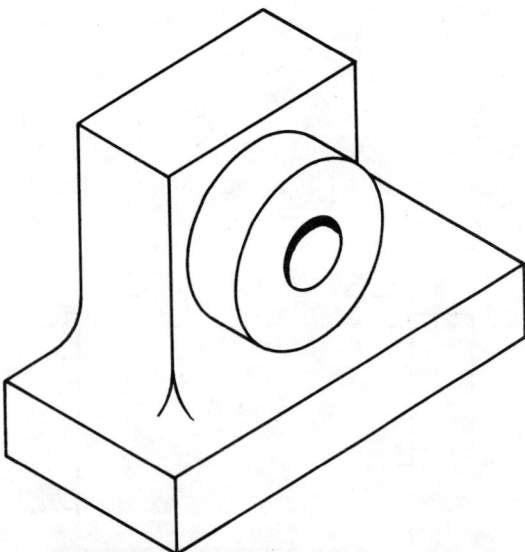

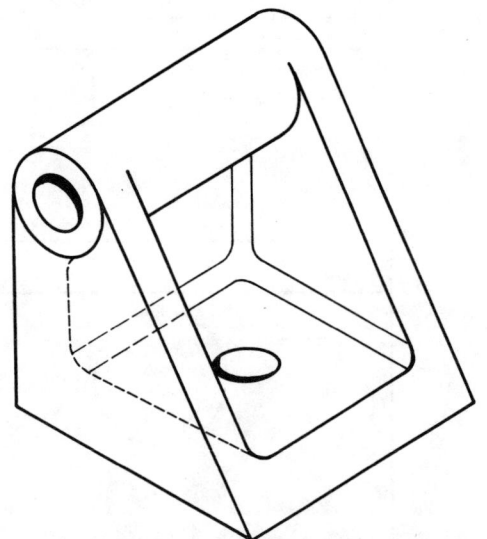

25 rad.

2 holes
10 dia.

Fillet rads. 6

25

12

25
dia.

30

40

12 rad.

10

10

40

70

12

40

70

50

10

70

50

2

50

2

35r

10r

35r

5r

35r

18

30

ø30

ø4
3 places
on 55 p.c.d

ø22
3 places

40 a/f

ø75

ø30

110

36

200

200

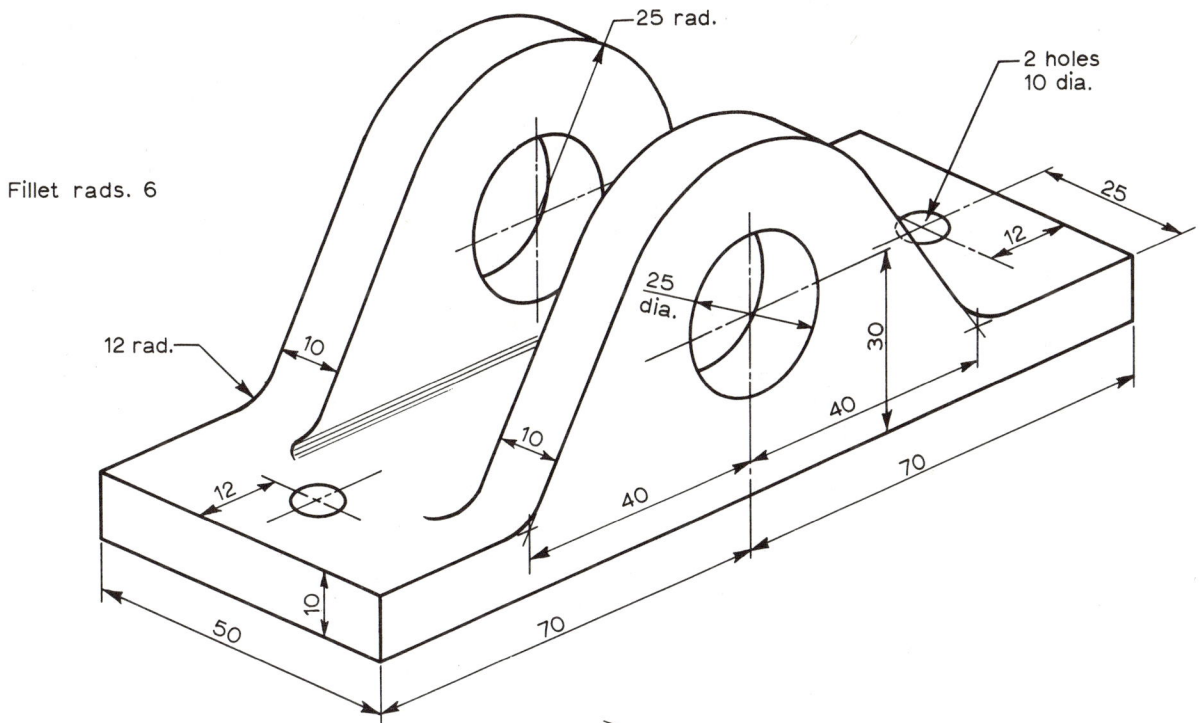

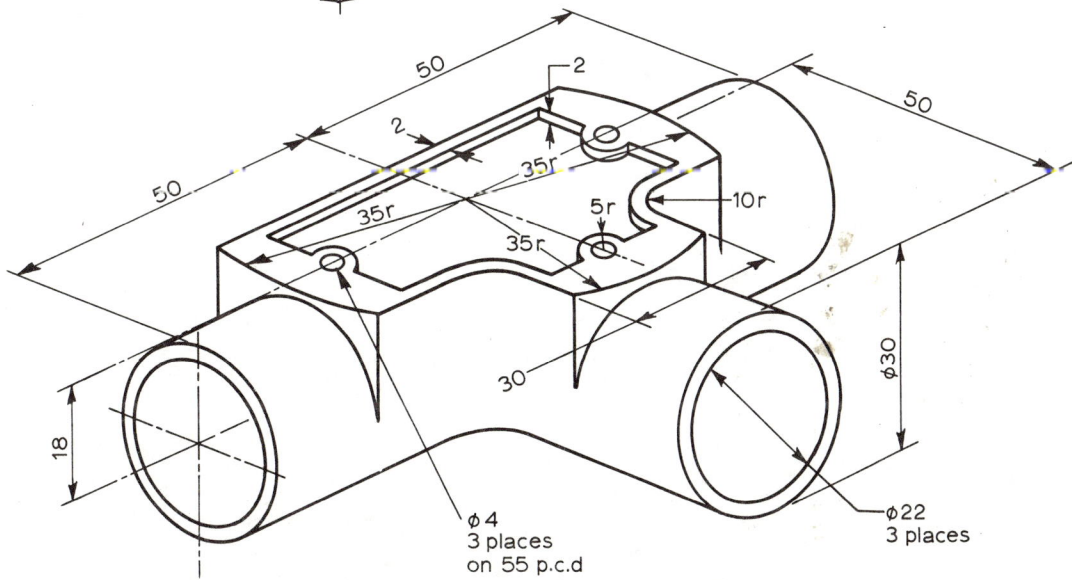

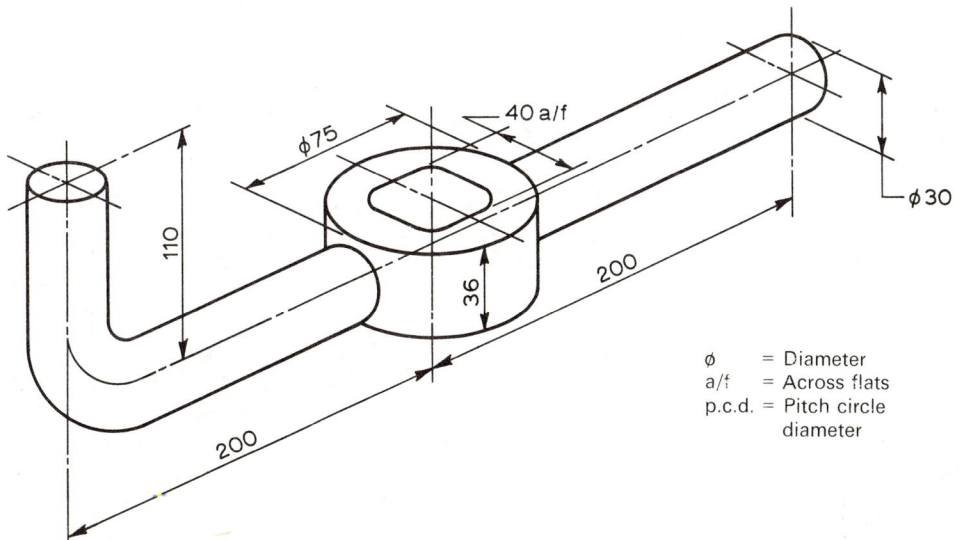

ø	= Diameter
a/f	= Across flats
p.c.d.	= Pitch circle
	diameter

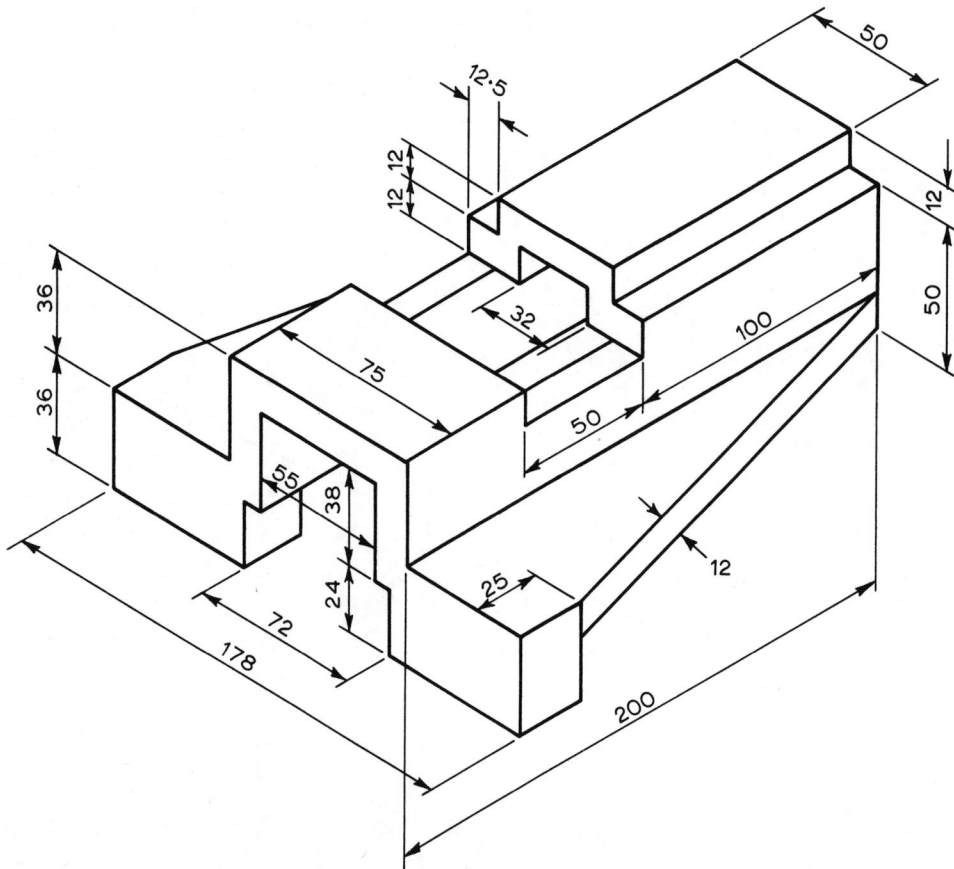

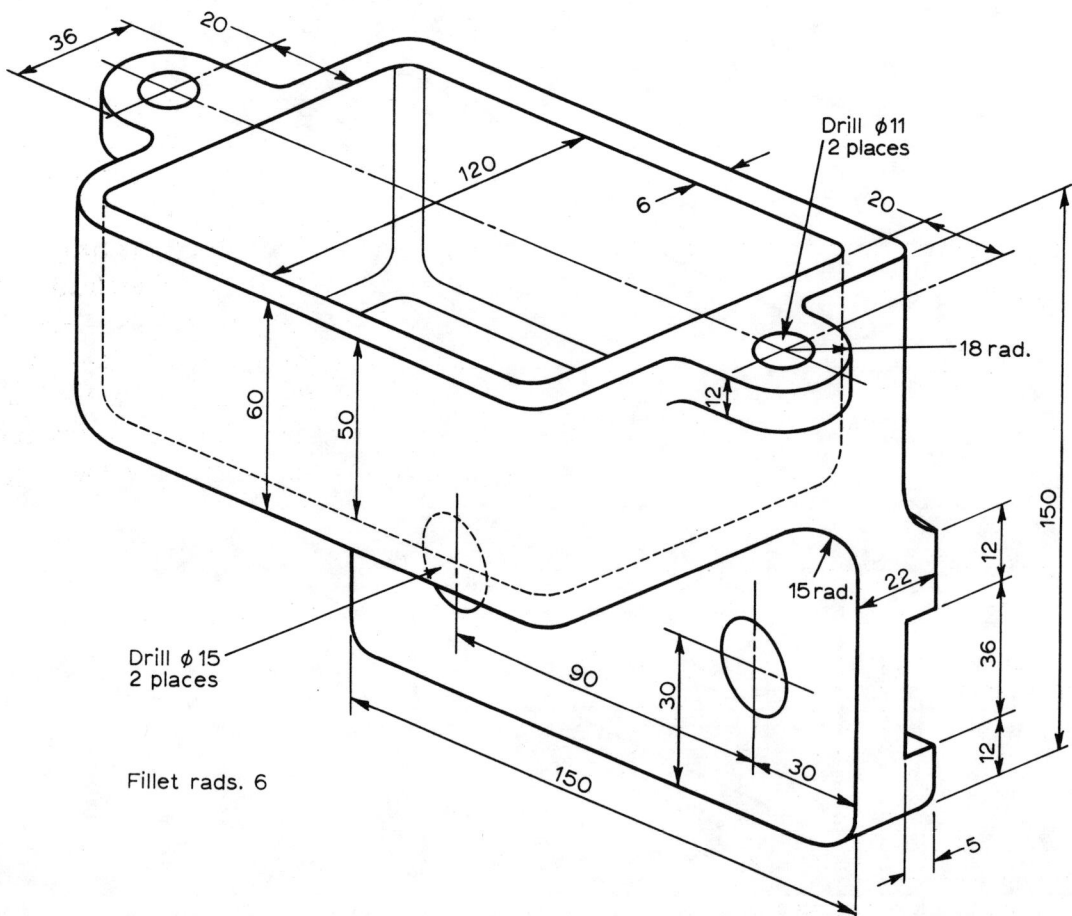

Drill φ11
2 places

Drill φ15
2 places

Fillet rads. 6

15 rad.

18 rad.

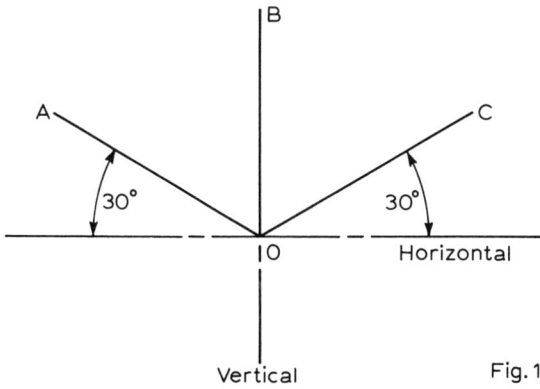

Fig. 1

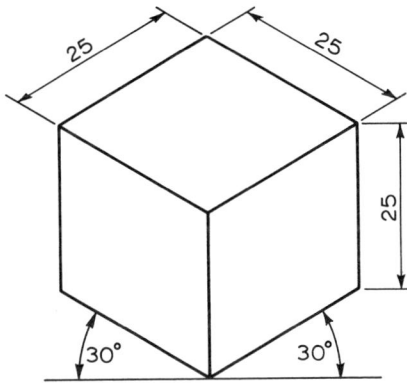

Fig. 2

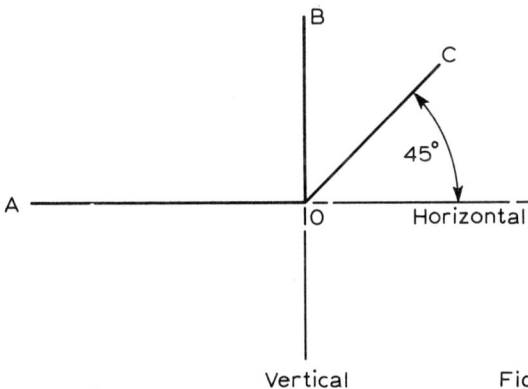

Fig. 3

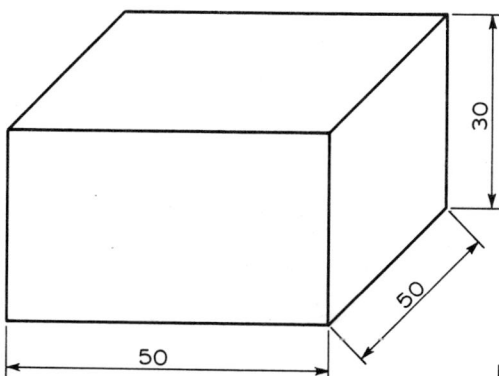

Fig. 4

Isometric and Oblique Drawings

From the preceding pages it will be seen that an *orthographic projection* is used to show views of individual faces. The number of views required depends upon the details of the component. This method is the most commonly used for production engineering, as it is often the only way to give the detailed information required to enable components to be manufactured correctly, direct from the drawings. These drawings may take a long time to prepare and there are times when sufficient information can be given on one drawing, especially for very simple details. For this purpose, we can use *isometric* or *oblique* projections, but the main purpose of these two projections is to provide a pictorial drawing of an object, which will convey shape and comparative size, rather than detailed dimensions.

An *isometric* projection shows both sides of an object to equal advantage (Fig. 2), whereas an *oblique* projection gives prominence to the side shown at the front (Fig. 4).

When constructing an isometric projection, the three axes are positioned as at Fig. 1 and all lines represent actual dimensions.

When constructing an oblique projection, the axes are as at Fig. 3. All vertical and horizontal lines are drawn to represent the actual dimensions, but the dimensions represented by the lines parallel to O–C are drawn to the ratio 1:2, i.e. a 25 mm line represents a 50 mm dimension (see Fig. 4).

Isometric or oblique circles can be obtained by transferring points of intersection from a full face drawing to the required projection.

First, a true square is drawn equal in width to the diameter of the required circle. The square is then divided into a convenient number of equal spaces by drawing vertical lines; these are then numbered. A similar square is drawn in the isometric view and the vertical lines are numbered as before. By transferring the distance between the centre line and the points of intersection on the true square, to the same numbered line on the projected face and joining the points by a series of arcs, the ellipse can be completed (see following illustration).

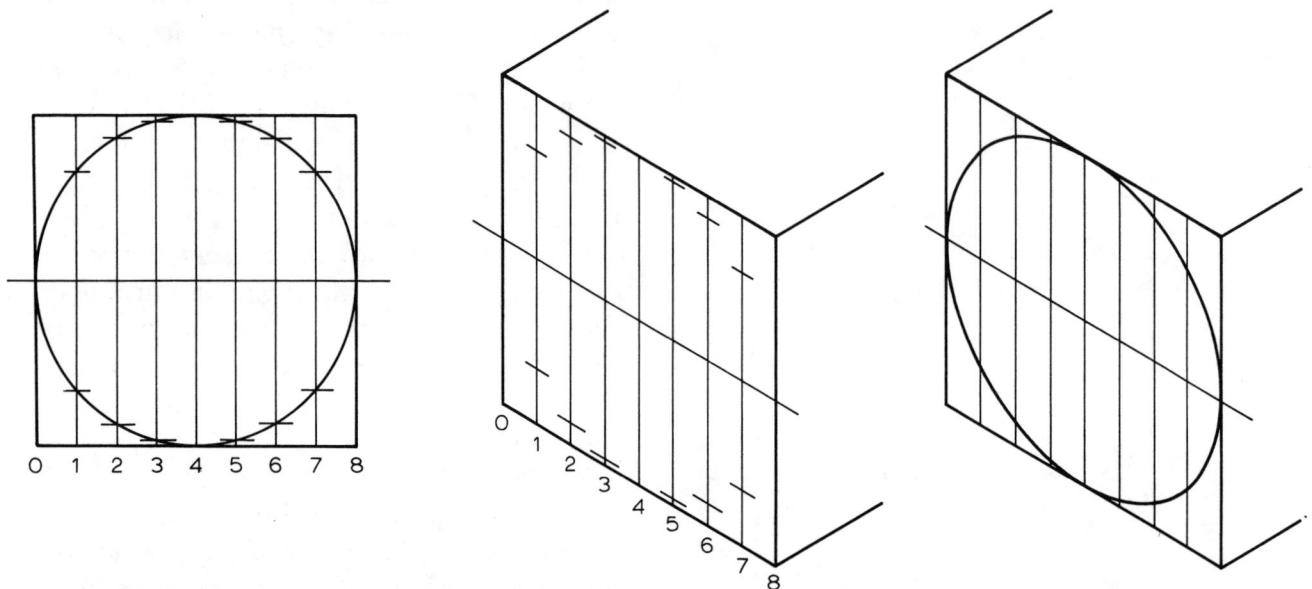

An alternative method which produces an approximate ellipse is shown in the illustration below.

First draw the isometric square equal to the diameter of the circle to be projected, then from point 1 draw line 1–3 and from point 4 draw line 4–5. Both these lines are drawn horizontally.

From point 1 draw line 1–6 and from point 4 draw line 4–2. Both these lines lay at 30 degrees from the vertical axis.

We now have points 1, 4, 7 and 8 from which to draw arcs, which will blend at points 2, 3, 5 and 6 and complete the ellipse.

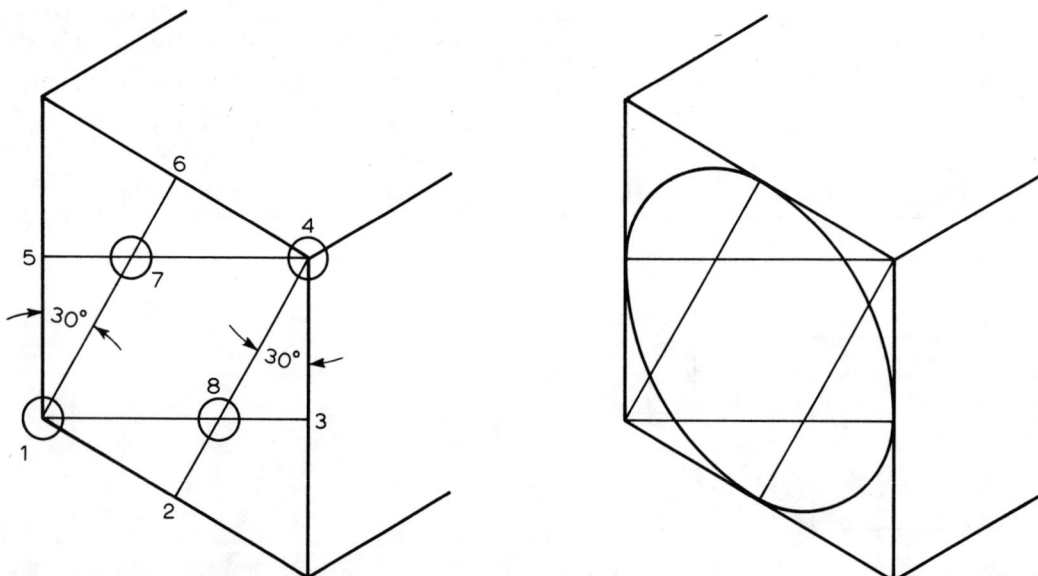

Sectional Views

Whilst some hidden detail can be shown on a drawing by using broken lines, more precise information can be given by drawing what would be seen if we were able to cut the component through to show the hidden detail. An imaginary cut can be made in any plane but there are rules to be followed when drawing what is known as a sectioned view.

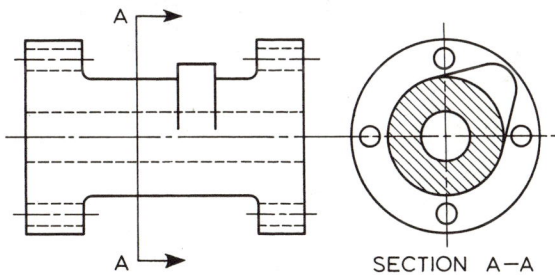

First, the drawing must indicate which face is being viewed. This is done by placing arrows at the extreme ends of the line that represents the imaginary cut.

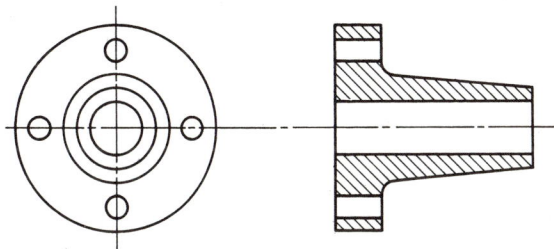

If the views of the detail being shown are identical in both directions the arrows may be omitted.

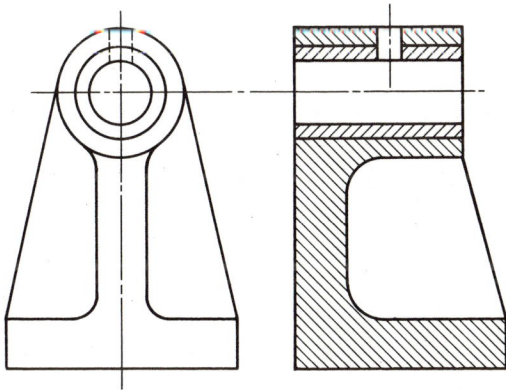

Secondly, the drawing must show whether the section line cuts through material or fresh air. This is done by hatching lines being drawn across any solid faces, leaving holes, bores, key-ways, etc., clear.
Where two or more surfaces are in contact in the sectioned view the hatching lines must be changed, either in direction, pitch, or slant-angle at the boundaries.

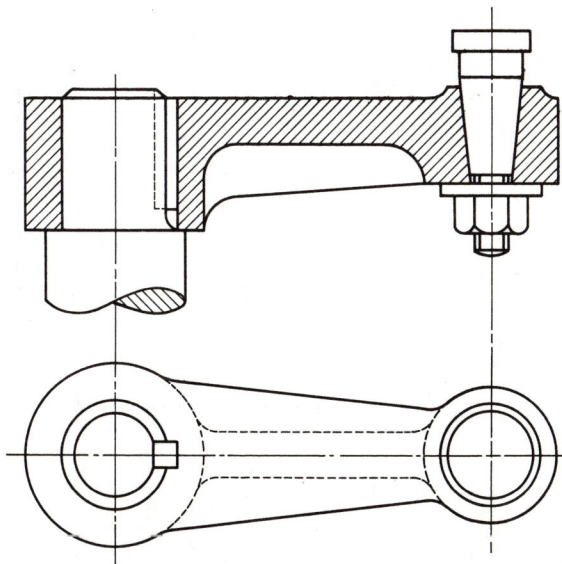

For the purpose of keeping a sectioned view clear and meaningful, such details as nuts, bolts, pins, solid shafts and webs of castings, etc., are not shown hatched, but are shown in outline only.

SECTION A—A

Section Views

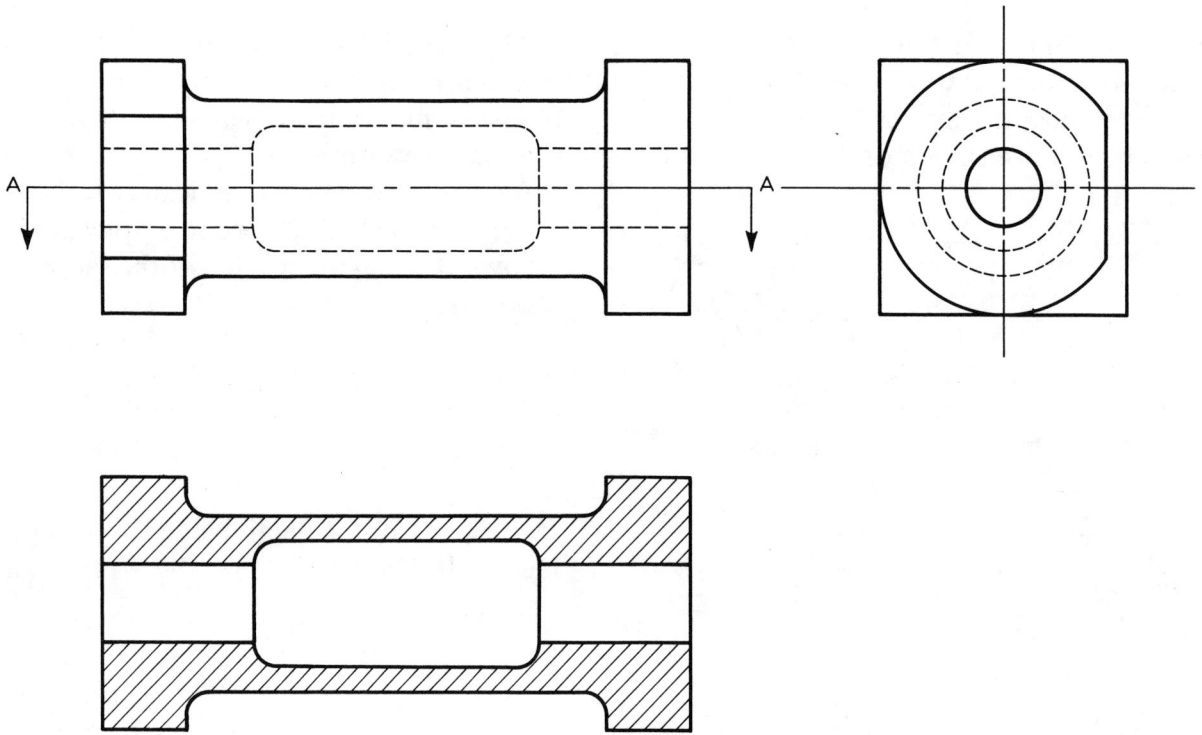

SECTION A — A

From the following drawing produce the sectioned view indicated X-X

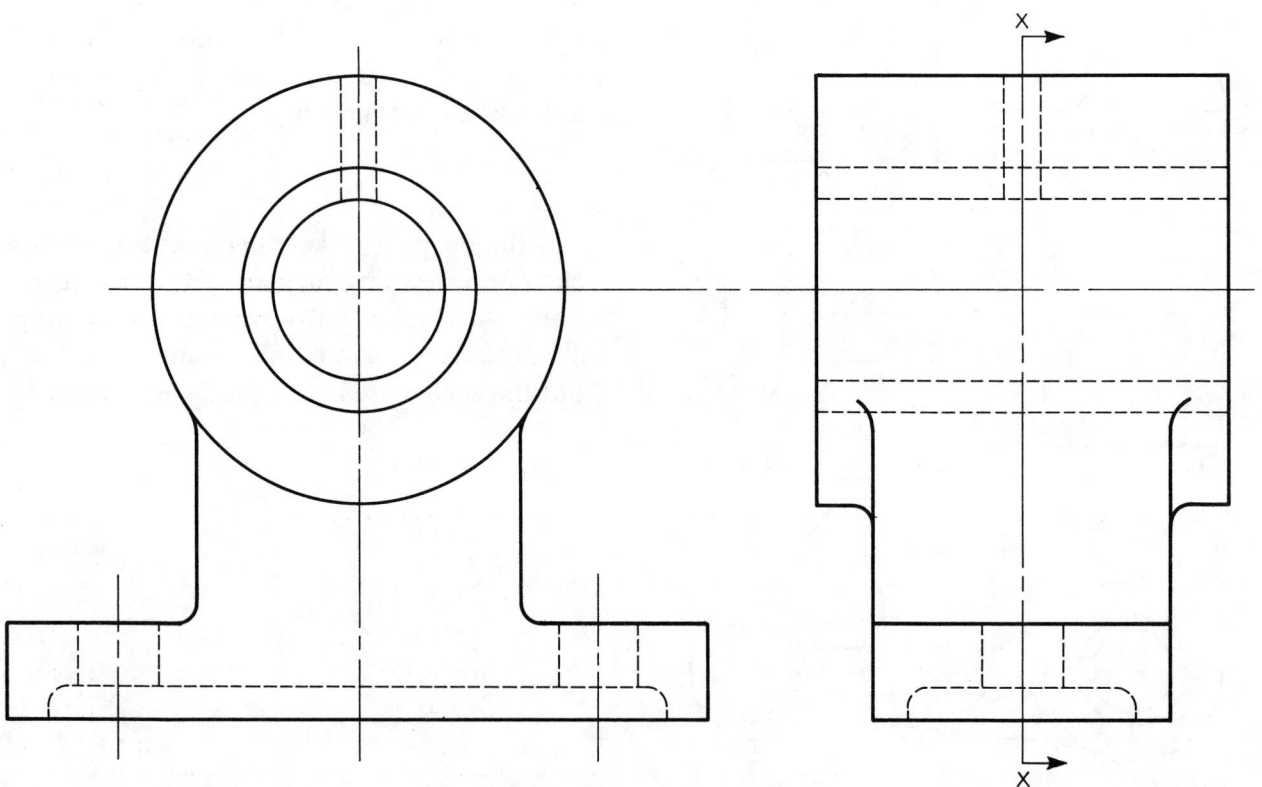

X

X

D3   Standard Conventions   D3

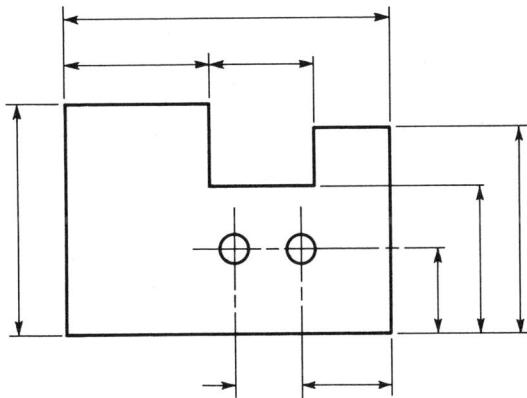

Dimensioning of Drawings

Dimension lines should be thin full lines and where possible placed outside the outline of the object (left). The purpose of this is to allow the outline of the object to stand out prominently from the other lines and to prevent any confusion between dimension lines and boundary lines. Dimension lines terminate in an arrow which should touch the line to which it refers. All dimensions should be placed in such a way that they can be read from the bottom right-hand corner of the drawing.

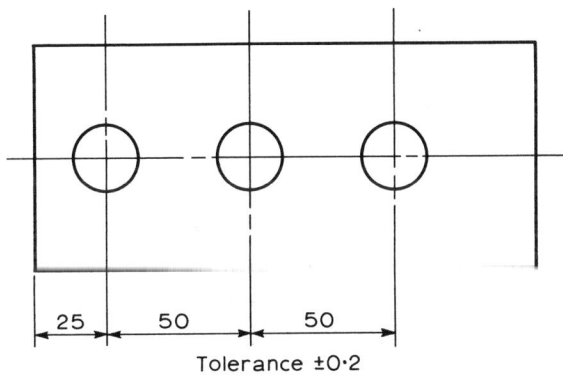

Where a number of dimensions are to be shown from consecutive features it is sometimes desirable to establish a datum from which to take several dimensions. This helps the production and inspection of a component but its main advantage lies in the fact that it eliminates the risk of a build-up of tolerances which might prove unacceptable at the time of assembly. Refer to the diagrams opposite and compare the position of the holes in the two plates when the holes are located at the extremes of positional tolerance.

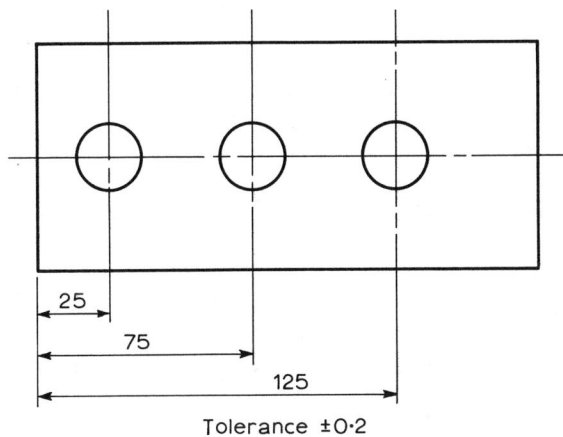

Tolerance ±0·2

Tolerance ±0·2

Apart from nominal sizes being shown on a drawing, the drawing must also indicate the limit of size to which the component must be produced, and here again with clarity of the drawing in mind, a reference table is often used to cover the majority of toleranced dimensions leaving plenty of space for special information.

A typical reference chart is shown below.

Tolerances except where otherwise stated		
Size		Tolerance
Up to 50 mm		±0·2 mm
Over 50 mm	Up to 150 mm	±0·4 mm
Over 150 mm	Up to 300 mm	±0·6 mm
Over 300 mm		±1 mm
On angles		±2°

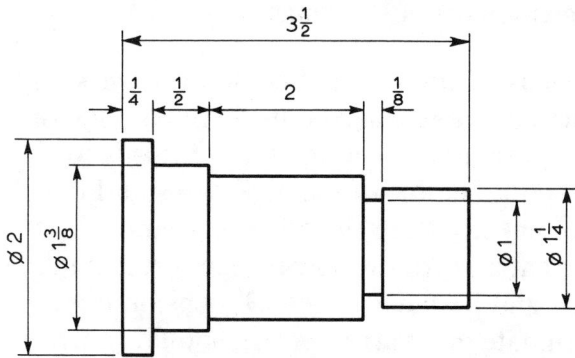

OVERALL DIMENSIONS PLACED OUTSIDE
INTERMEDIATE DIMENSIONS

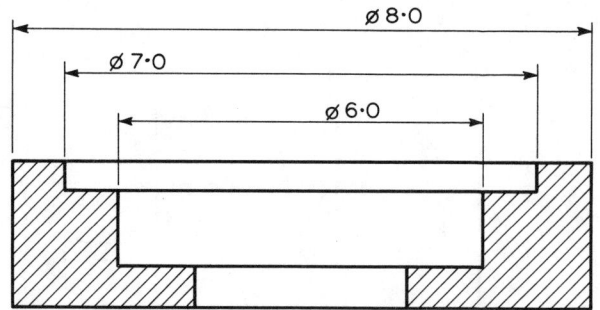

DIMENSIONS STAGGERED TO AVOID CONFUSION

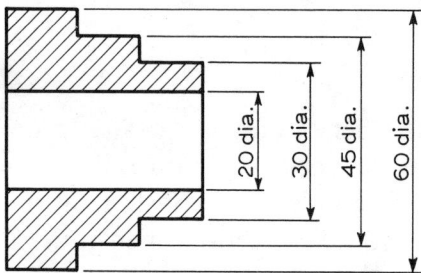

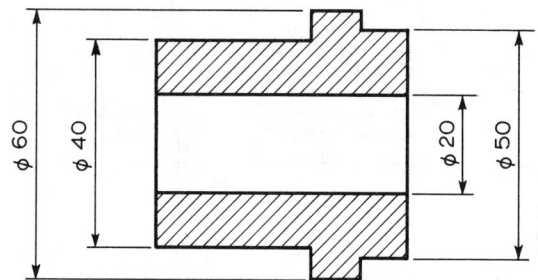

ALTERNATIVE METHODS OF INDICATING DIAMETERS

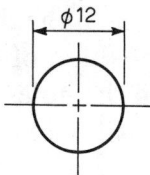

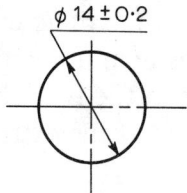

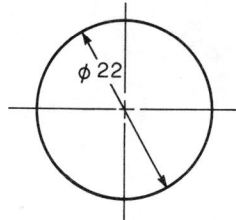

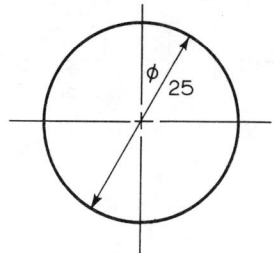

DIMENSIONING CIRCLES AND ANGLES

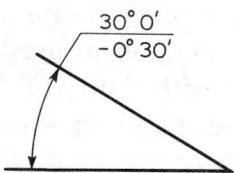

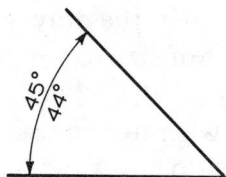

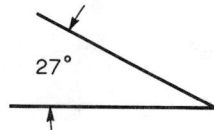

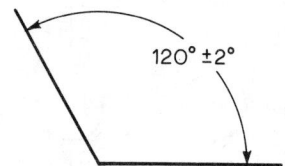

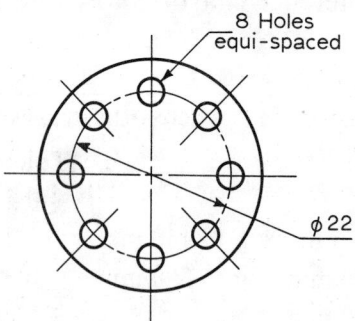

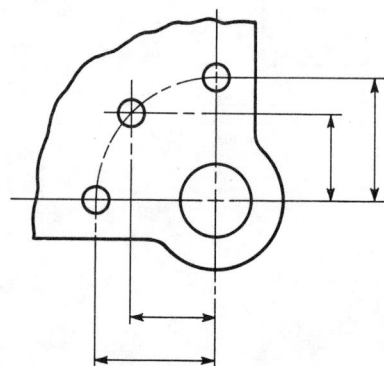

POSITIONING OF HOLES

64

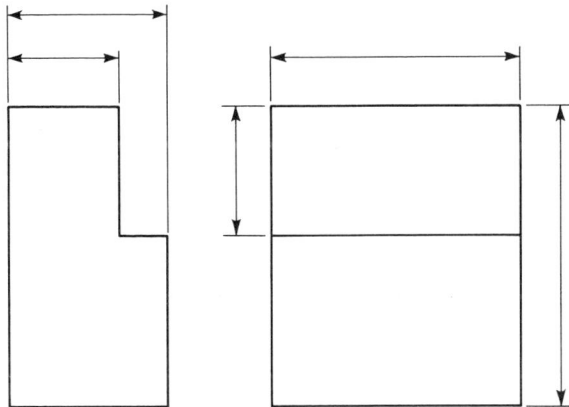

Indication of Machined Surfaces

Apart from a drawing showing the shape and size of an object, it is important that the surface finish required and the means of achieving it be indicated.

Consideration of surface finish is very important in two ways: One, in that if the surface finish is not good enough the component may fail in operation. An example of this can be found in hydraulic systems, where valve and piston surfaces must be of a very high standard to prevent loss of pressure or oil. Another example of super surface finish can be found on slip gauges, where firstly, without a super finish they could not be guaranteed to the required accuracy, and secondly, they could not be wrung together.

On the other hand these super finishes are very costly to produce and it would be very undesirable to waste time and money producing them where they serve no useful purpose, i.e. where the surface merely controls a dimension and has no contact with another surface or where the surface is finally painted for protection. Here quite rough surfaces are often permissible, even to the point of leaving a cast surface unmachined.

The fact that a surface must be machined is denoted on the drawing by a closed tick thus ∇, and the required surface finish to be obtained is indicated by a number placed on the closed tick thus ∇. These numbers refer to micro-inches when used with Imperial dimensions (1 micro-inch = 0·000 001 inch), or microns when used with metric dimensions (1 micron = 0·001 mm).

Note: the lower the number the higher the class of surface finish.

When it is important to denote that a particular surface finish is to be maintained and *not* machined, the symbol and note is used ∇. *Do not machine,* and where it is immaterial whether or not the surface is to be machined, provided that the surface requirements are met, the symbol ∇ is used. Note that in both cases an open tick is used.

Note. ∇ all over

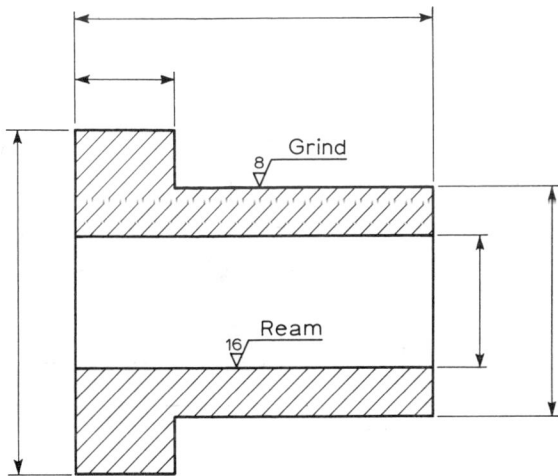

$$\frac{125}{\nabla}$$ all over except where stated

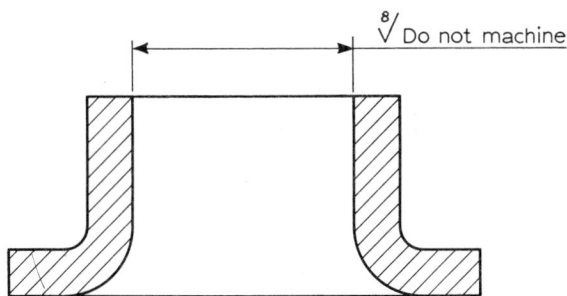

Grind
8 ∇

Ream
16 ∇

8 ∇ Do not machine

65

Welding Symbols

The scheme of symbols used to indicate the type, size and position of welds on drawings is laid down in B.S. 499, Part 2 (1965), from which the following symbols have been extracted.

Type of Weld	Symbol	Section Through Weld
Fillet		
Square butt		
Single vee butt		
Double vee butt		
Single U butt		
Double U butt		
Single bevel butt		
Double bevel butt		
Single J butt		
Double J butt		

To indicate the position of the weld an arrowhead is used to show the reference side.
The side nearest the arrow is called the arrow side and the opposite side is called the other side.
The reference line connected to the arrowhead is the datum for determining the position of the
weld symbol.

1   When welds are to be made from the arrow
    side of the joint: symbol is turned upside
    down and suspended from the
    reference line.

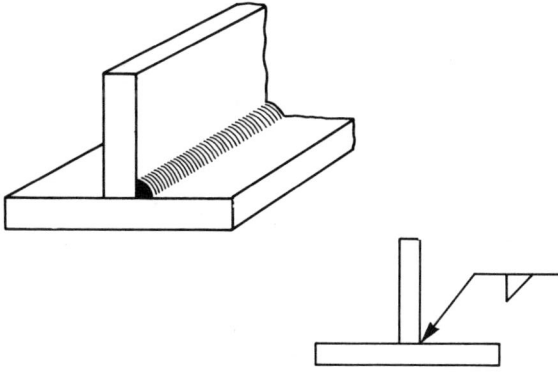

2   When welds are to be made from the other
    side of the joint: symbol is placed on
    the reference line.

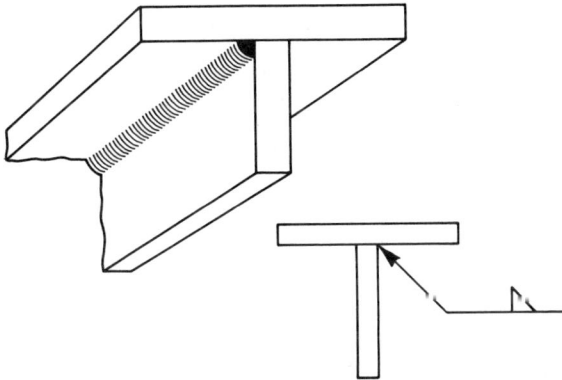

3   When welds are to be made on both sides
    of the joint: symbol is disposed either side
    of the reference line.

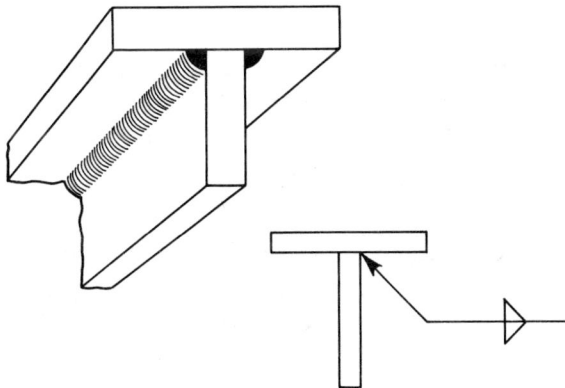

4   When edge preparation occurs on only
    one of the plates to be joined, the arrow
    points to that plate which is to be prepared,
    the symbol indicating the direction in
    which the edge preparation takes place.

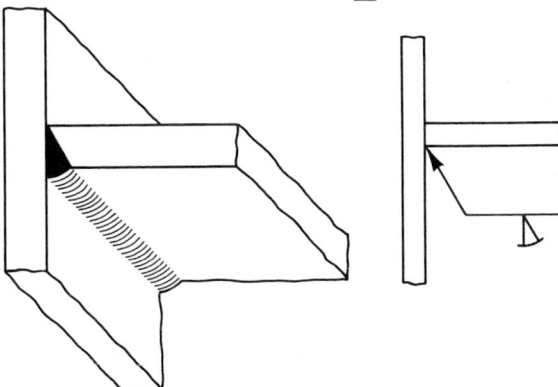

The symbols shown above and on the previous page are a selection of the types likely to be met at
this stage of training. Reference should be made to B.S. 499, Part 2 (1965) for the complete
scheme of symbols used.

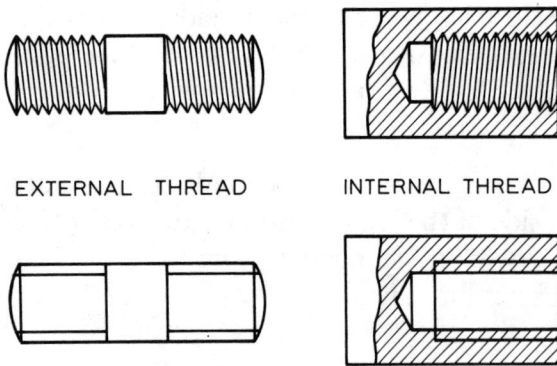

EXTERNAL THREAD

INTERNAL THREAD

CONVENTIONAL METHOD OF
REPRESENTATION

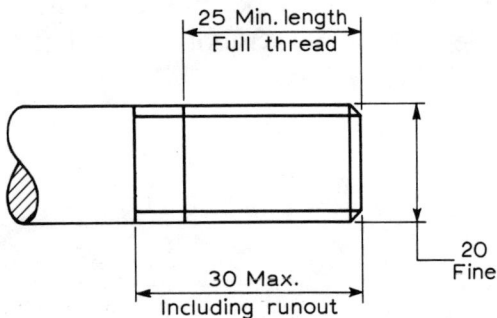

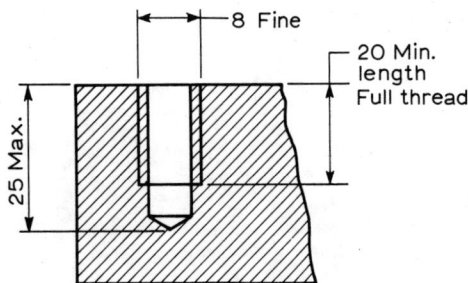

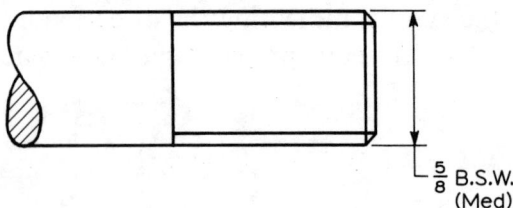

25 Min. length
Full thread

20
Fine

30 Max.
Including runout

8 Fine

20 Min.
length
Full thread

25 Max.

$\frac{5}{8}$ B.S.W.
(Med)

Conventional Representation of Screw Threads

To draw a screw thread in full detail is a very tedious job and in most cases serves no useful purpose. All information required can be given in note form. Convenient methods of representing threaded parts have been devised and these should be used where possible. When specifying special threads, the major, minor, and effective diameters should be stated, along with the form of threads required. The limits of size must also be stated.

The majority of threads conform to an accepted standard or system, and by stating the outside diameter of the thread and the system required, all the relevant information can be obtained from reference charts.

With the British Association series, each thread has a designating number to which reference is made.

Standard metric threads are either of the Fine series or the Coarse series, and the series must be stated along with the diameter.

The following abbreviations should be used when drawing threads of a standard series:

B.S.W.	for British Standard Whitworth
B.S.F.	„ British Standard Fine
B.S.P.	„ British Standard Pipe
B.S.Con.	„ British Standard Conduit
B.A.	„ British Association
U.N.C.	„ Unified Coarse
U.N.F.	„ Unified Fine
S.I.	„ Système Internationale

D4   Limits and Fits

A limit is the amount of variation in size a designer allows, taking into consideration its effect on the final assembly. Two limits are required for each dimension, the *top limit* and the *bottom limit*. The difference between these two limits is called the *tolerance*, i.e. tolerance = top limit − bottom limit.

It is virtually impossible to produce components exactly to size and a tolerance has to be allowed. The smaller this tolerance the more expensive it becomes to produce the component and the designer and production engineer have to decide whether to use a small tolerance, more expensive component or a larger tolerance component which costs less to produce.

The aim of the production engineer is to ensure the production of mechanically sound units at competitive prices; therefore, people employed on production must make full use of any tolerance allowed, as they are there to assist in speedy production.

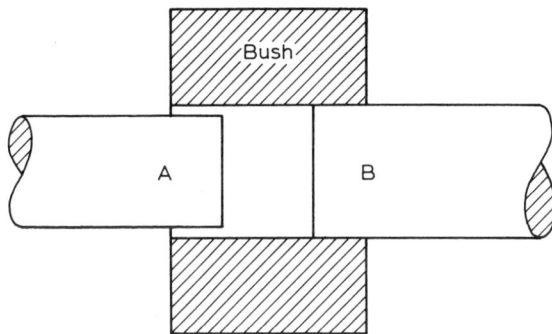

Mechanical assemblies involve various types of fit:

If a shaft is required to revolve inside a bush, there must be clearance between the shaft and the bush. This type of fit is called a *clearance fit* (as at A).

If the two components are to be forced together such that they become as one component then this type of fit is called an *interference fit* (as at B).

Transition fit is the name given to the type of fit that cannot be classified as one of the above, i.e. both components are the same size.

Methods of Indicating Limits on Drawings

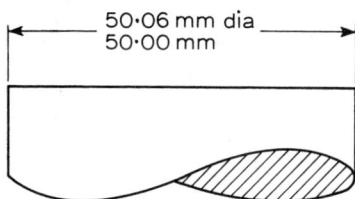

Top limit   = 50·06 mm
Bottom limit = 50·00 mm
Tolerance   =   0·06 mm

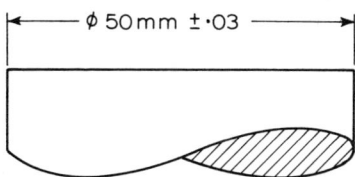

Top limit   = 50 mm + ·03 = 50·03 mm
Bottom limit = 50 mm − ·03 = 49·97 mm
Tolerance   =   0·06 mm

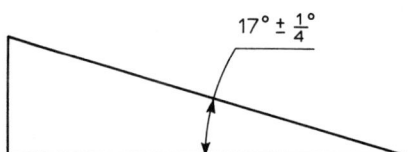

Top limit   = $17° + \frac{1}{4}° = 17\frac{1}{4}°$
Bottom limit = $17° - \frac{1}{4}° = 16\frac{3}{4}°$
Tolerance   = $\frac{1}{2}°$

Gauges

Gauges are used for checking dimensions to ensure that only those that are within the required limits are accepted.

A gauge cannot measure any particular size, but by using two gauges, one made to the top limit and the other to the bottom limit, any components that are outside their prescribed limits can be detected. This type of gauge is called a *limit gauge*.

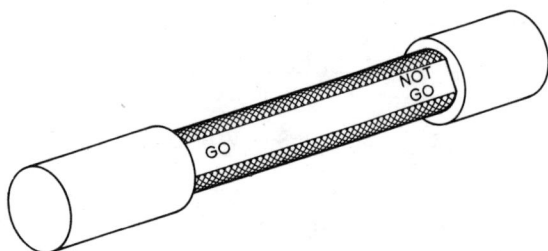

DOUBLE ENDED PLUG GAUGE

Plug Gauge

This type of gauge is used to check internal bores and holes. The bottom limit or GO end of a plug gauge is made longer than the NOT GO end because it is the end that is subjected to most wear because it must pass through every hole it checks. The NOT GO end should not enter the hole; if it does the component should be rejected.

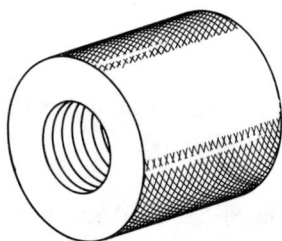

RING GAUGE

Ring Gauge

This type of gauge is used similarly to the above, but is used to check the external diameters of shafts, etc.

When using gauges to check tapered holes or shafts, only one plug or ring gauge is required. This is because they can be stepped to show the two required diameters.

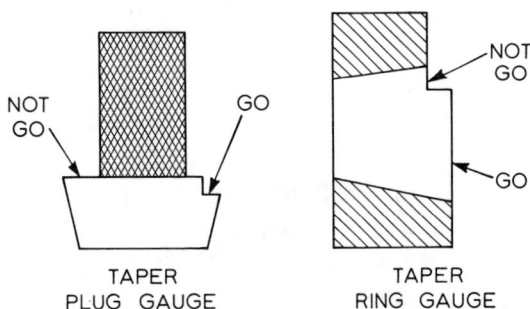

TAPER
PLUG GAUGE

TAPER
RING GAUGE

Gap Gauges

Gap gauges are often used where it is impossible to use a ring gauge, e.g. undercuts on shafts, circlip grooves, etc.

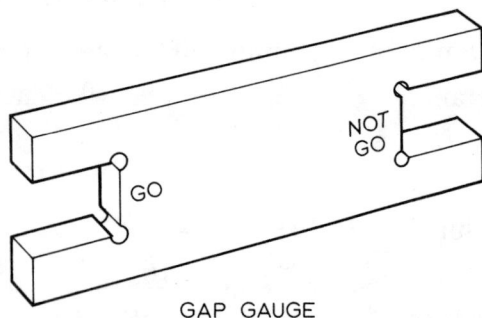

GAP GAUGE

Gauges are usually made of hardened and ground steel, but even so they still get worn and damaged and require periodic inspection.

For very accurate work, gauges are fitted with insulated handles to prevent the heat from the hand affecting the size of the gauge.

D   Test Questions   D

Section D2

1   What is the major difference between first and third angle projection?
2   What is the difference between isometric and oblique projection?
3   Sketch pictorially the following items:
    - (a)   Calipers;
    - (b)   Dividers;
    - (c)   Trammels;
    - (d)   Metric micrometer;
    - (e)   Metric vernier;
    - (f)   Engineer's square;
    - (g)   Vee block;
    - (h)   Angle plate;
    - (i)   Three turning tools;
    - (j)   Shaper tool;
    - (k)   Reamer;
    - (l)   Lock-nut;
    - (m)   Castellated nut;
    - (n)   Four types of rivet;
    - (o)   Depth gauge;
    - (p)   Soldering iron.
4   Sketch sectional views of the components shown on page 56.

Section D3

5   Sketch in third angle projection three views of each of the components shown on page 56 and include suitable dimensions.
6   What is meant by each of the following symbols?

7   What do the following abbreviations stand for:
    U.N.C., B.S.F., S.I., B.A., U.N.F., B.S.W.

Section D4

9   What are the three classes of fit?
10   Sketch a shaft and bearing indicating each class of fit.
11   What is meant by top limit, bottom limit and tolerance?
11   What is meant by GO and NOT GO when marked on a plug gauge?
12   Why is the GO end of a plug gauge longer than the NOT GO end?
13   For what are plug gauges, ring gauges and gap gauges used? Sketch each type in use.

E1   Mechanical Properties of Materials   E1

The Hazards of Drilling Holes in Prestressed Concrete and Structural Members

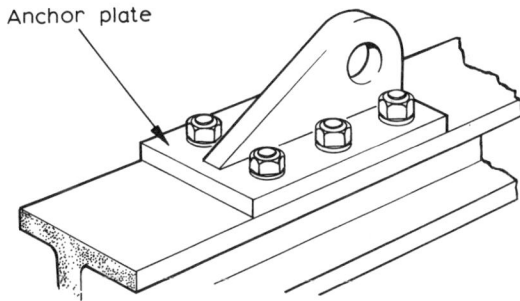

The erection and assembly of steelwork and other structures requires the use of all types of lifting tackle and winches. To secure these to the structure, as it progresses from the initial layout to its completion, stays and bracing are secured to pillars and beams. These securing points should not be bolted into the structure. Any holes drilled for bolting purposes will cause a reduction in the cross-sectional area of the section and may cause the pillar to bend or buckle at that point.

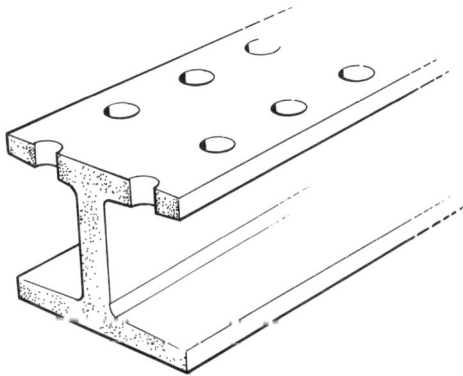

The Building Regulations which govern the assembly of structures, etc., regard this as a grave hazard to the life of the building and to the lives of people who work in it.

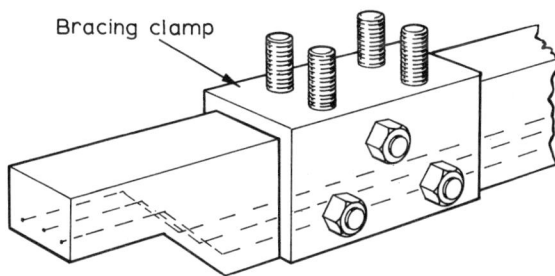

Reinforced and prestressed concrete beams and pillars have steel rods and bars cast into them to give greater strength and rigidity. If the bracing points or anchor points are bolted through them, then the rods may be cut by the drill and the concrete beam will fail when loaded.

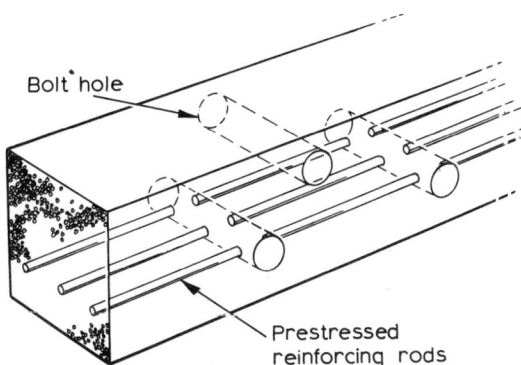

Prestressed beams will lose their initial built-in stress and will collapse if their reinforcing rods are cut by drilling.

E2   Mechanical Fastening

Riveted and Bolted Joints—Types, Identification and Defects

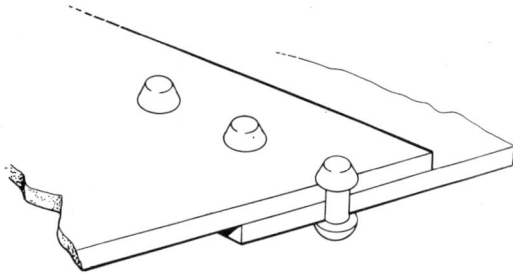

Single-riveted Lap Joint

The top illustration shows a single-riveted lap joint. It is the most common of all bolted and riveted joints.

It is used for joining two plates together with a simple overlap of each plate, secured with a single row of rivets or bolts.

The single-riveted lap joint is used on boilers, trunking, coachwork and sheet and thin metal joints.

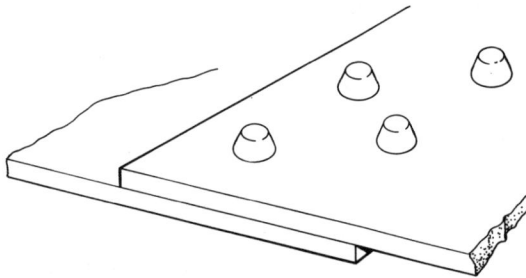

Double-riveted Lap Joint

Joints subjected to higher pressure and stresses, steam and gas tightness, require a larger overlap for two rows of staggered rivets (zig-zag riveting). This joint is called a double-riveted lap joint.

The double-riveted lap joint is used on low-pressure vessels, exhaust and ducting systems, and washing plant.

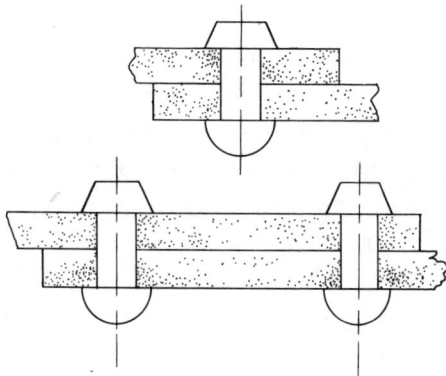

Assembly of Lap Joint

In the above joints the strength of the joint against failure is the resistance offered by one cross-sectional area of the rivets. The assembly of the joint, however, is fairly simple; a podging spanner and drifts are used to bring the holes into alignment.

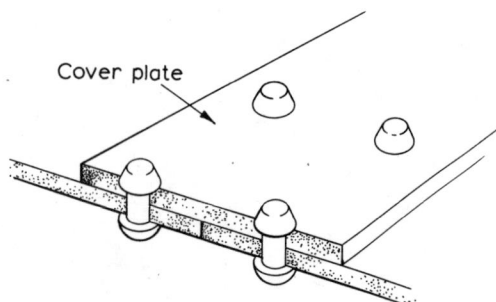

Cover plate

Single Cover-plate Butt Joint

Another adaptation of a lap joint is the butt joint with a single cover plate. The rivets may be single or double row, either side of the joint face. The main plates of the joint butt against one another, hence the name butt joint.

74

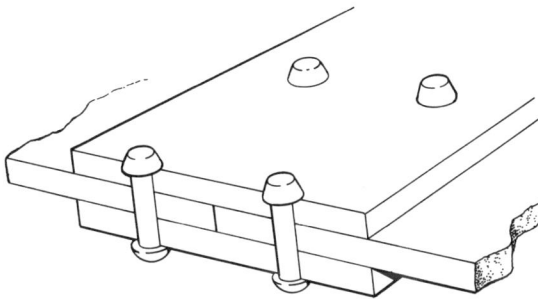

Double Cover-plate Butt Joint

A stronger resistance is offered to external forces by the following butt joints which have a double cross-sectional area of rivet to resist any pull on the joint (shown in illustration on left). Note that the main plates butt together, hence the name double cover-plate butt joint.

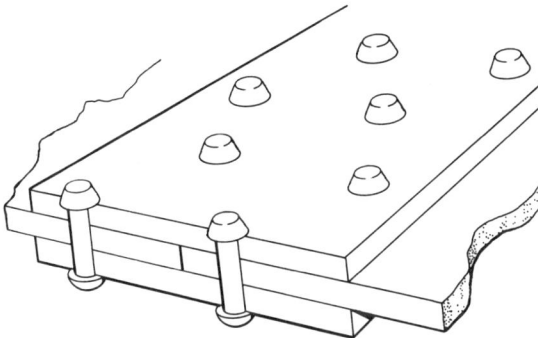

Double-riveted, Double Cover-plate Butt Joint

The joint may also be strengthened against excessive pressure by the insertion of further rows of rivets. These give an increase in the resisting cross-sectional area of the rivets. The illustration (left) shows a double-riveted, double cover-plate butt joint.

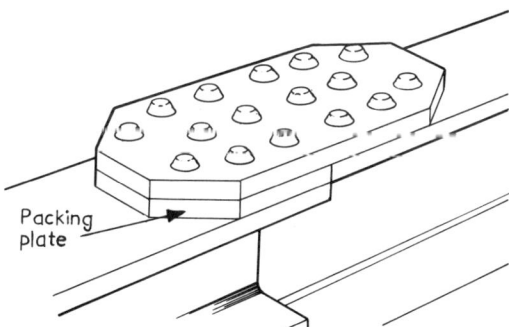

Splice Joint

When a column or pillar is reduced in section then a splice joint is used with packing and thrust plates inserted. This joint is usually riveted or bolted on site. The joists butt on to the thrust plate to allow the loading to be transferred through them rather than the rivets. Joints may be of plates overlapping— lap joints—or they may be of plates butting together—butt joints. They may have separate plates at top and bottom (cover plates), single or double, and they may have as many rows of rivets as is required.

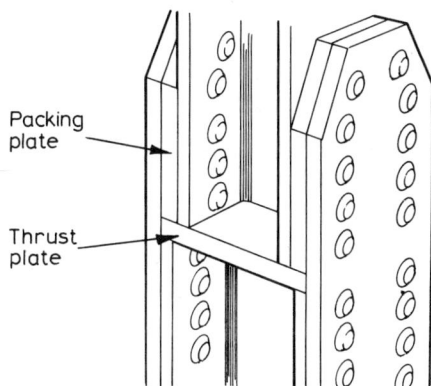

Packing plate

Packing plate

Thrust plate

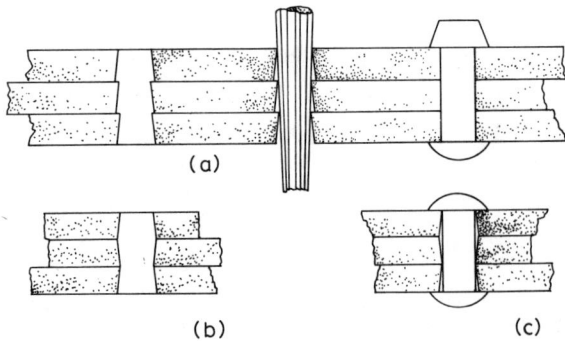

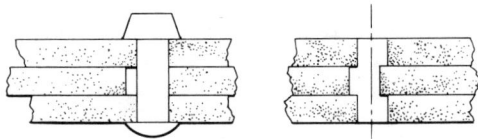

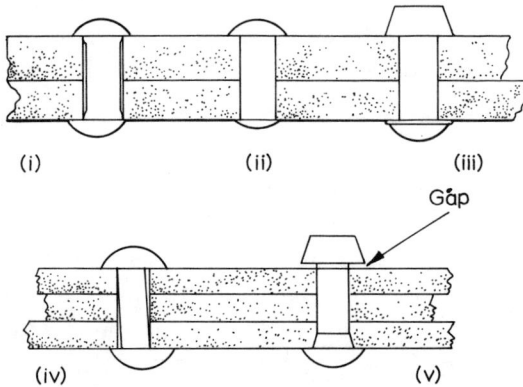

Defects of Riveted Joints

Riveted joints can fail because of poor quality or workmanship of the plater or riveter. On assembly, components with punched holes in them must be placed together, as shown by (a) in the top illustration, and then a correct-sized tapered reamer is passed through them to clear the sharp edges off the hole and make it parallel and to size along its length. When assembled as at (b) and (c), the rivet will not fill the hole and the joint may fail.

During assembly, the over-use of reamers and tapered drifts can cause oblique holes which again may not be filled by the rivet metal when it is closed in the hole.

The main causes of failure of riveted joints are:

1 Overheated and burnt rivets.
2 Rivets being too cold—they will not fill the hole and a correct head is not formed; see (i) in illustration on left.
3 Rivets used which are too short and are incorrect—a weak head is formed; see (ii).
4 Too long a rivet causes a misshapen head called a 'jockey cap'; see (iii).
5 Rivets driven into holes obliquely with head and tail not in line; see (iv).
6 Rivets not 'held up' correctly during closure by the use of a hand-riveting gun —will rattle when tapped with a light hammer; see (v).

Other failures may be due to laminated plate —the plate tears from rivet hole to edge of plate.

E3   Screw Threads   E3

B.S.W., B.S.F., B.S.P., B.A., U.N.F. and U.N.C. Threads

The British Standard Whitworth series was the result, in 1841, of the efforts made by Sir Joseph Whitworth to get accepted some form of standardization of screw threads. The thread is of the vee form. The angle is 55 degrees and it is rounded at the root and the crest. The B.S.W. series is used in all branches of general engineering.

Extract from the British Standard Whitworth Series

Nominal Size (in)	Threads/ Inch	Thread Depth (in)	Effective Diameter (in)	Thread Radius (in)	Tapping Size (mm)
$\frac{1}{4}$	20	0·032	0·218	0·007	5·10
$\frac{5}{16}$	18	0·036	0·277	0·008	6·50
$\frac{3}{8}$	16	0·040	0·335	0·009	7·90
$\frac{7}{16}$	14	0·046	0·392	0·010	9·25
$\frac{1}{2}$	12	0·053	0·447	0·011	10·50
$\frac{5}{8}$	11	0·058	0·567	0·012	13·50
$\frac{3}{4}$	10	0·064	0·686	0·014	16·25
1	8	0·080	0·920	0·017	22·00

The British Standard *fine* series was introduced by the Engineering Standards Committee, in 1908, to supplement the Whitworth series in cases where finer pitches were required. The thread is the same as the Whitworth Standard.

Extract from the British Standard Fine Series

Nominal Size (in)	Threads/ Inch	Thread Depth (in)	Effective Diameter (in)	Thread Radius (in)	Tapping Size (mm)
$\frac{1}{4}$	26	0·025	0·225	0·005	5·30
$\frac{5}{16}$	22	0·029	0·283	0·006	6·75
$\frac{3}{8}$	20	0·032	0·343	0·007	8·25
$\frac{7}{16}$	18	0·036	0·402	0·008	9·70
$\frac{1}{2}$	16	0·040	0·460	0·009	11·10
$\frac{5}{8}$	14	0·046	0·579	0·010	14·00
$\frac{3}{4}$	12	0·053	0·697	0·011	16·75
1	10	0·064	0·936	0·014	22·75

The British Standard *pipe* series is the standard used when threading pipes for gas, water or steam. The thread form is the same as the Whitworth Standard. The nominal size refers to the bore of the pipe. From the tables it will be seen that this series is finer than the others, allowing greater locking power and also preventing excessive weakening of the pipe wall.

Extract from the British Standard Pipe Series (parallel)

Nominal Size (in)	Major Diameter (in)	Threads/ Inch	Thread Depth (in)	Effective Diameter (in)	Thread Radius (in)	Tapping Size (mm)
$\frac{1}{8}$	0·383	28	0·023	0·360	0·005	8·75
$\frac{1}{4}$	0·518	19	0·033	0·484	0·007	11·80
$\frac{3}{8}$	0·656	19	0·033	0·622	0·007	15·25
$\frac{1}{2}$	0·825	14	0·046	0·779	0·010	19·00
$\frac{5}{8}$	0·902	14	0·046	0·856	0·010	21·00
$\frac{3}{4}$	1·041	14	0·046	0·995	0·010	24·50
1	1·309	11	0·058	1·251	0·012	30·75
$1\frac{1}{4}$	1·650	11	0·058	1·592	0·012	39·50

The British Association series is recommended for use below $\frac{1}{4}$ in (6·35 mm) dia. The vee thread has an angle of $47\frac{1}{2}$ degrees and is rounded top and bottom. It is based on the Swiss national series, which explains why all the dimensions are in a metric unit (millimetres). This series is extensively used throughout light engineering and instrument manufacture.

Extract from the British Association Series (all sizes in mm)

B.A. No.	Major Diameter	Pitch	Thread Depth	Effective Diameter	Thread Radius	Tapping Size
0	6·00	1·00	0·60	5·40	0·18	5·10
2	4·7	0·81	0·49	4·21	0·15	4·00
4	3·6	0·66	0·40	3·20	0·13	3·00
6	2·8	0·53	0·32	2·48	0·10	2·30
8	2·2	0·43	0·26	1·94	0·075	1·80
10	1·7	0·35	0·21	1·49	0·05	1·40
12	1·3	0·28	0·17	1·13	0·05	1·05

The Unified series (U.N. *fine* and U.N. *coarse*) were introduced to obtain some standardization between Great Britain, the U.S.A. and Canada. They are a compromise between the American National Series and the British Whitworth Series.

The thread angle is 60 degrees. The bolt is rounded at the root and the crest, but the nut is rounded at the root only.

Extract from the Unified Fine Series (U.N.F.)

Nominal Size (in)	Threads/ Inch	Thread Depth (in)	Effective Diameter (in)	Tapping Size (mm)
$\frac{1}{4}$	28	0·022	0·227	5·50
$\frac{5}{16}$	24	0·026	0·285	6·90
$\frac{3}{8}$	24	0·026	0·348	8·50
$\frac{7}{16}$	20	0·031	0·405	9·90
$\frac{1}{2}$	20	0·031	0·467	11·40
$\frac{5}{8}$	18	0·034	0·589	14·50
$\frac{3}{4}$	16	0·038	0·709	17·50
1	12	0·051	0·946	23·25

Extract from the Unified Coarse Series (U.N.C.)

Nominal Size (in)	Threads/ Inch	Thread Depth (in)	Effective Diameter (in)	Tapping Size (mm)
$\frac{1}{4}$	20	0·031	0·217	5·20
$\frac{5}{16}$	18	0·034	0·276	6·60
$\frac{3}{8}$	16	0·038	0·334	8·00
$\frac{7}{16}$	14	0·044	0·391	9·40
$\frac{1}{2}$	13	0·047	0·450	10·80
$\frac{5}{8}$	11	0·056	0·566	13·50
$\frac{3}{4}$	10	0·061	0·685	16·50
1	8	0·077	0·919	22·25

Section E Fabrication and Welding

E3   **E3**

British Standard Conduit Thread

The *British Standard Conduit* thread series is, as the name implies, the one used when threading conduit and fittings for electric wiring purposes.
The form of the thread is the Whitworth Standard.

Note: The nominal size refers to the outside diameter of the conduit.

		British Standard Conduit Thread Series (dimensions in inches)							
		Nominal size of conduit							
		$\frac{1}{2}$	$\frac{5}{8}$	$\frac{3}{4}$	1	$1\frac{1}{4}$	$1\frac{1}{2}$	2	$2\frac{1}{2}$
Threads per inch		18	18	16	16	16	14	14	14
Depth of thread		0·036	0·036	0·040	0·040	0·040	0·046	0·046	0·046
Length of threaded ends of conduit	min.	$\frac{3}{8}$	$\frac{7}{16}$	$\frac{1}{2}$	$\frac{5}{8}$	$\frac{11}{16}$	$\frac{3}{4}$	$\frac{7}{8}$	1
	max.	$\frac{7}{16}$	$\frac{1}{2}$	$\frac{9}{16}$	$\frac{11}{16}$	$\frac{3}{4}$	$\frac{13}{16}$	$\frac{15}{16}$	$1\frac{1}{16}$
Length of threaded portions of fittings	min.	$\frac{7}{16}$	$\frac{1}{2}$	$\frac{9}{16}$	$\frac{11}{16}$	$\frac{3}{4}$	$\frac{13}{16}$	$\frac{15}{16}$	$1\frac{1}{16}$
	max.	$\frac{1}{2}$	$\frac{9}{16}$	$\frac{5}{8}$	$\frac{3}{4}$	$\frac{13}{16}$	$\frac{7}{8}$	1	$1\frac{1}{8}$
Locknuts { Thickness		$\frac{3}{16}$	$\frac{3}{16}$	$\frac{1}{4}$	$\frac{1}{4}$	$\frac{1}{4}$	$\frac{5}{16}$	$\frac{3}{8}$	$\frac{3}{8}$
Width across flats		0·815	0·915	1·092	1·472	1·662	2·040	2·570	3·150

80

E4   Sheet Metal Work and Fabrication   E4

Stiffening of Fabricated Material

Stiffening for Strength

An important aspect of design is to ensure that a component will not bend when a load is applied. This can often be achieved by increasing the thickness of the material, but if rigidity is increased by this method the finished product will weigh and cost more, due to the extra material used.

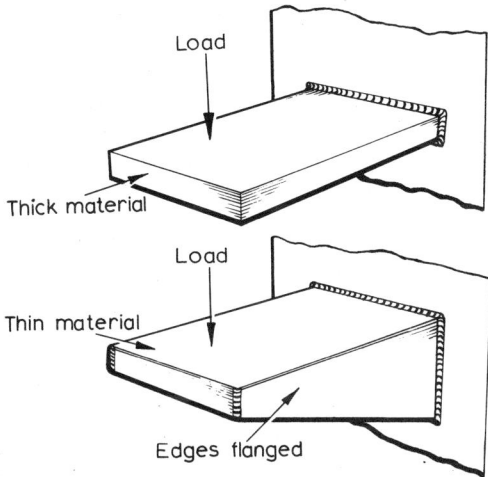

The illustration opposite shows a bracket welded to a plate. If the bracket was designed in this way, even only to withstand a fairly light loading, it would have to be made of reasonably thick material.

The second illustration shows the bracket redesigned from thinner material. The flanges have been folded, resulting in a strong bracket with a great saving in material and weight.

Stiffening for Appearance and Safety

There are many ways in which sheet materials can be stiffened. It should be remembered that the stiffening is not only for strength reasons but also for appearance and safety. The edges of thin sheets can be extremely dangerous, causing cuts due to sharp-pointed corners, etc. Thin sheets should never be left, after fabrication, with a raw edge.

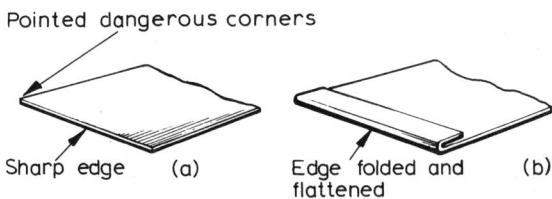

Illustrations (a) and (b), on the left, show a simple method of folding the edge of a sheet to make it safe. The appearance of the folded edge is far more pleasing to the eye than a raw edge.

A wired edge, as shown opposite, combines rigidity, appearance and safety.

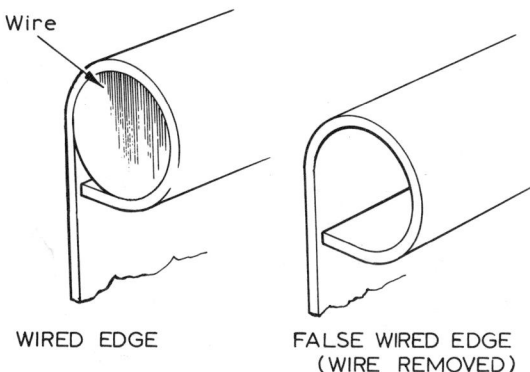

It is impractical to use wires of large diameter. For example, in order to obtain rigidity on a sheet of thin material 2 m long it would be necessary to use a wire of about 25 mm diameter. This would obviously be costly and heavy. In this situation a false wired edge would be used, as shown opposite. The edge is formed around the wire which is then withdrawn leaving a tube section formed from the sheet.

Achievement of Strength by Using Welded or Riveted Angle Frames to Form Stiffeners

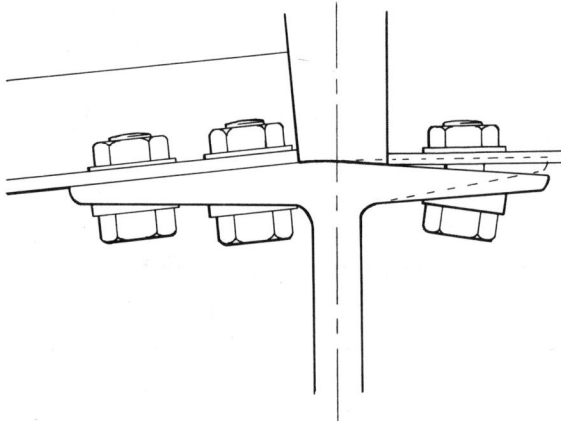

A beam which is to take heavy loading will require a flange. The weight may cause the flange to bend, so to avoid such bending, stiffeners made from plate and angle are formed to fit snugly under the flange.

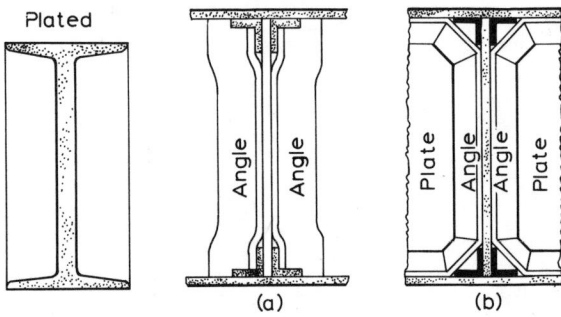

The stiffeners may be of various forms and are either forged (joggled)—see (a) in the second illustration—or cut and welded (kneed), (b), to fit between the flanges. They are placed at regular intervals along the length of the beam or at precise load points.

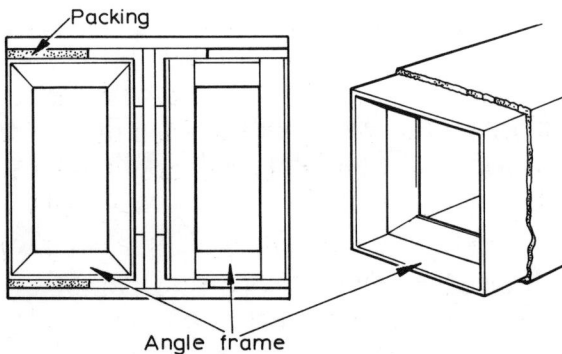

When the beam flange is wide in comparison with its depth, then angle frames of box section are fitted under the flanges, or the beam is designed as a series of angle frames linked together by plates on all four sides to form a box beam.

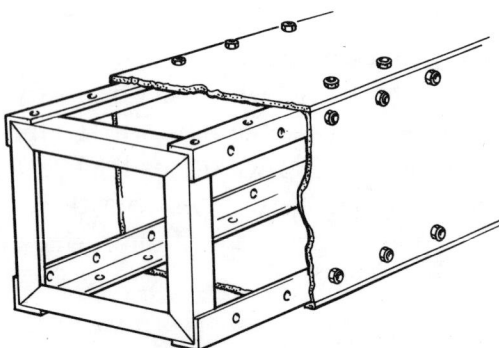

This type is used, for lightness and strength, in bridges, tall structures, and dockside and overhead cranes. A further increase in strength is achieved by joining the angle frames with angles at each corner and the whole plated as shown in the bottom illustration.

E5   Joining of Materials E5

Joining Conduits

Conduit lengths form a strong and continuous system of protection for electrical conductors. The separate lengths are joined together by couplings which are supplied with the internal thread cut in them. The illustration below shows a coupling and a length of conduit.

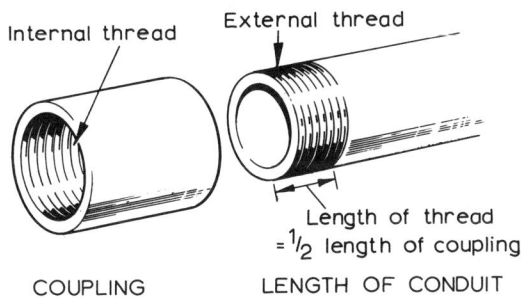

Internal thread — External thread

Length of thread = $\frac{1}{2}$ length of coupling

COUPLING   LENGTH OF CONDUIT

Preparing the Conduit

1   The conduit is cut to length, keeping the end face square.
2   The bore of the conduit should be reamered to remove any sharp edges. This is to prevent damage occurring to the wires when they are drawn in.
3   Cutting compound should be smeared on the outside of the tube and stocks and dies used to cut the external thread. The dies should be reversed every few forward turns to break the cuttings and clear the thread.

Joining With a Running Coupling

In many situations it is not possible to rotate the conduit to enable it to be screwed into a fitting or coupling. Instead, a running coupling joint is made. The two lengths of conduit are joined together with a coupling and secured with a locknut. The procedure for making the joint is as follows:

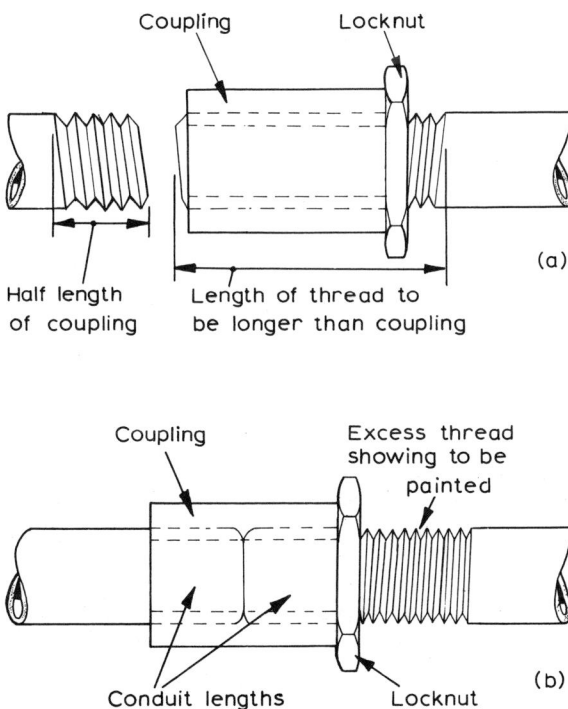

Coupling   Locknut

Half length of coupling   Length of thread to be longer than coupling   (a)

Coupling   Excess thread showing to be painted

Conduit lengths   Locknut   (b)

1   The external thread on one of the conduits is made longer than the coupling. The locknut and coupling are then screwed on to this conduit—see illustration (a).
2   The two lengths of conduit are then butted together and the coupling screwed back on to the second conduit. When the coupling is up tight, the locknut is screwed up to the coupling to prevent it turning—see illustration (b).
3   The excess thread showing on the first conduit is then painted to prevent rusting.

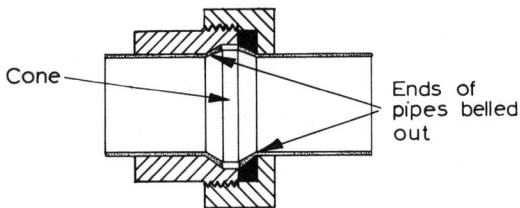

Jointing of Pipes

Compression Joints

1   Manipulative joint suitable for soft copper pipes, etc.

2   Non-manipulative joint suitable for copper pipes, etc., up to 50 mm bore.

3   Suitable for plastic tubing. Note the liner which fits inside the two tubes, preventing them from being crushed.

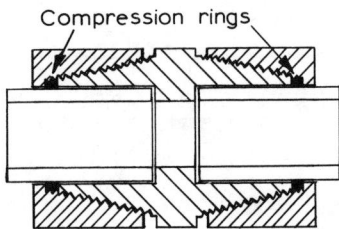

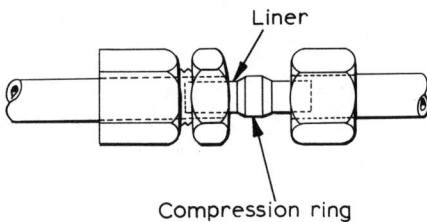

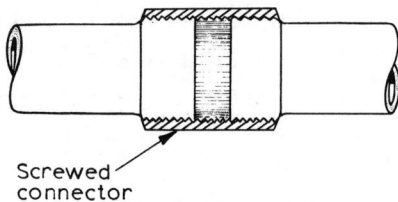

Screwed Joints

1   A binding material is usually wound into each of the male threads to effect the seal.

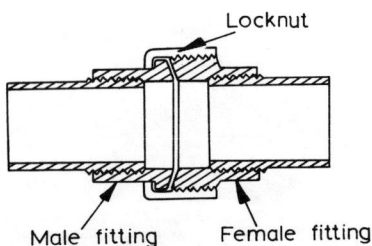

2   This type of fitting enables easy removal of portions of piping in a system. The joint is normally made with a jointing compound.

Flanged Joints

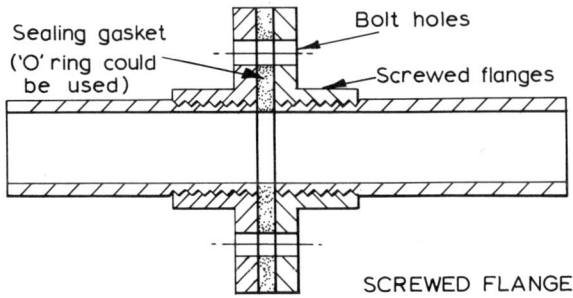

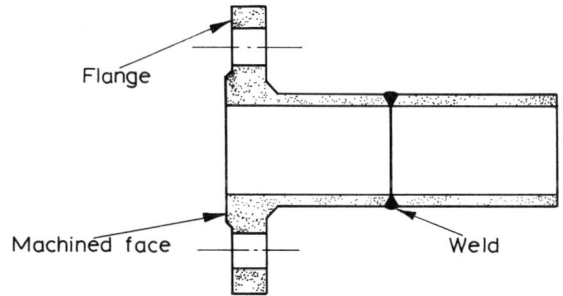

SCREWED FLANGE

With the type of welded flange shown on the left, it is necessary to machine the face after welding to remove splatter and smooth the final run.

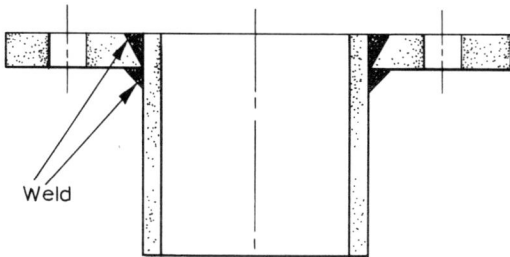

Hose Connectors

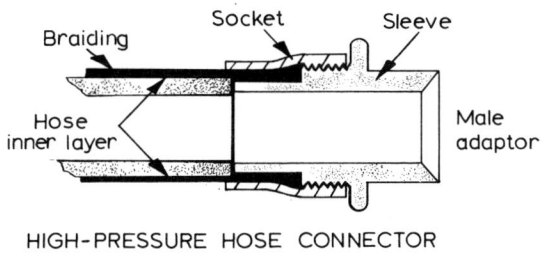

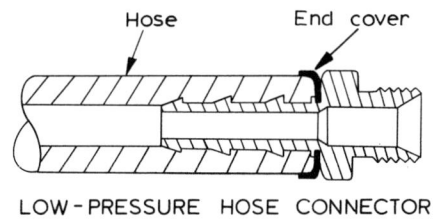

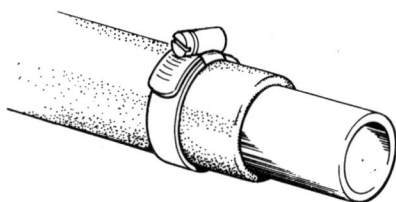

HIGH-PRESSURE HOSE CONNECTOR

LOW-PRESSURE HOSE CONNECTOR

A worm-drive clip, of the type shown on the left, may be used to connect flexible hose to a metal pipe.

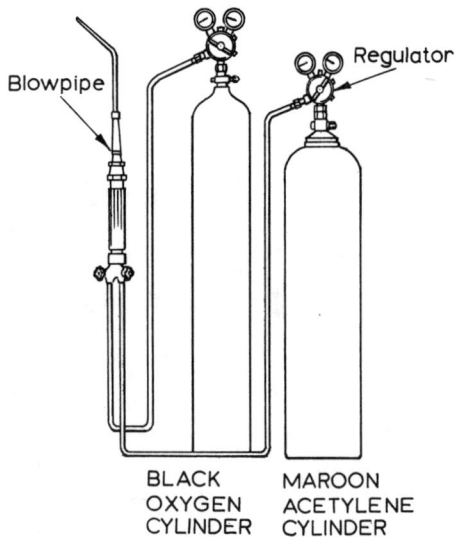

Blowpipe
Regulator
BLACK OXYGEN CYLINDER   MAROON ACETYLENE CYLINDER

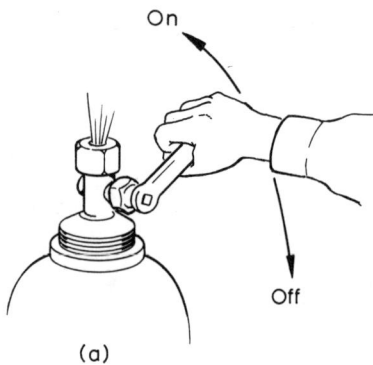

On
Off
(a)

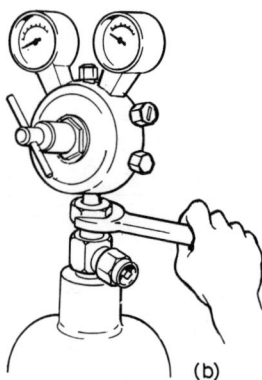

(b)

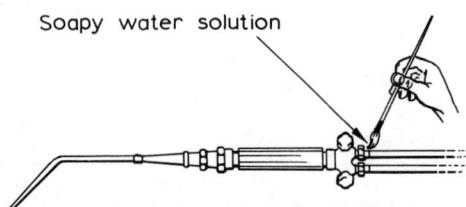

Soapy water solution

Oxy-Acetylene Welding

Equipment

A complete oxy-acetylene welding set consists of the following:

1   An oxygen cylinder; 2   An oxygen regulator; 3   A black oxygen hose; 4   An acetylene cylinder; 5   An acetylene regulator; 6   A red acetylene hose; 7   A blowpipe and set of interchangeable nozzles or tips; 8   A trolley to carry the cylinders; 9   Keys and spanners to fit the equipment; 10   Filler rods and fluxes; 11   Personal protection for the operator, i.e. gloves, goggles, apron, etc.

Assembly of the Equipment

1   Clean the valve seat on the cylinders by cracking open the valve momentarily (this will remove any dust or grit which would damage the valve seat), as shown in (a) on the left.
2   Select the correct regulator and screw it into the cylinder valve seat until it is finger-tight. Then, using the correct spanner, tighten up to obtain a gas-tight joint, as shown in (b). Slacken the pressure-control screw to relieve the pressure on the regulator diaphragm.

Never use oil or grease on the threads.

3   If new hoses are to be used they should be blown through to remove any dust or chalk.
4   Connect the hoses to the regulators—*Black* hose for *oxygen. Red* hose for *acetylene.* Make sure that the anti-flashback devices are fitted to the blowpipe end of the hoses.
5   Arrange the hoses so that they are not twisted.
6   Connect the hoses to the correct connectors on the blowpipe.
7   Open the blowpipe valves and turn on the gas supplies.
8   Adjust the regulators until the correct pressures are obtained. When these pressures have been set, the blowpipe valves are closed.
9   The system should now be tested for leaks, using a soapy water solution—see left.

Leftward Welding

In this method of welding, the blowpipe should be grasped firmly, ensuring that the wrist is free to move. The weld is commenced on the *right-hand* side of the seam, working towards the left-hand side, as shown in illustration (a) below. The blowpipe is moved forward with the flame pointing in the direction of the welding, with the filler rod being held in front of the flame. The angles of inclination of the blowpipe and filler rod are shown clearly in illustration (b). The blowpipe is given small sideways movements, while the filler rod is moved steadily across the weld seam.

The filler rod metal is added using a backward and forward movement of the rod, allowing the flame to melt the bottom edges of the plates just ahead of the weld pool. It is important that the filler rod is not held continuously in contact with the weld pool, or the heat from the flame cannot reach the bottom edges of the joint.

The disadvantages of this method are that the view of the joint edges is interrupted and it is necessary to remove the end of the rod, which slows down the welding process. Also, as the rod is removed, oxides form on the butt end which are then deposited into the weld pool when recommencing the weld.

For welding plates over 3·2 mm ($\frac{1}{8}$ in), it is usual to bevel the edges to approximately 40°, giving an included angle of approximately 80°. With plates over 8·2 mm ($\frac{5}{16}$ in), it is usually necessary to use two or more passes. When welding plates 6·5 mm ($\frac{1}{4}$ in) thick and over it is difficult to obtain complete firmness and even penetration at the bottom of the vee and if the welder is not highly skilled the quality of the weld decreases as the plate thickness increases. Also, with an 80° vee the amount of filler metal used is quite considerable and can lead to problems due to the expansion and contraction that is produced. For these reasons it is considered that the maximum economical thickness of metal for this technique is 5 mm ($\frac{3}{16}$ in).

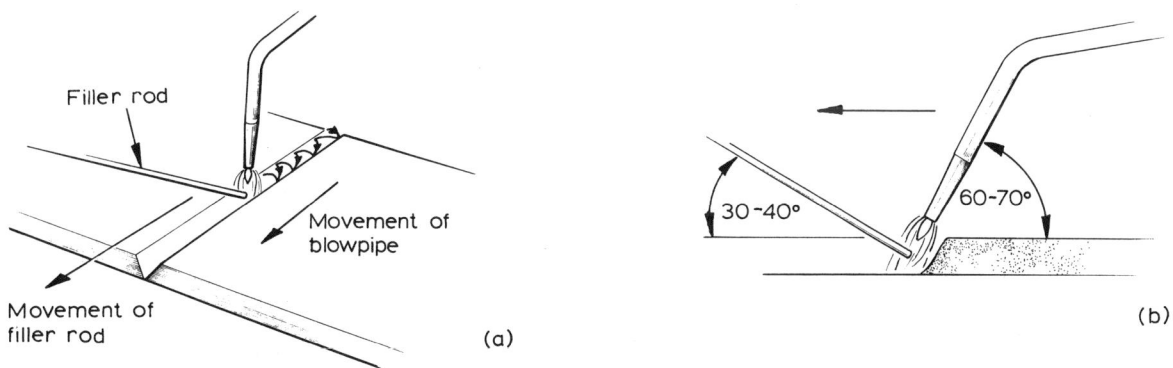

(a)

(b)

Leftward Technique

Thickness of Plate		Power of Blowpipe (gas consumption)		Diameter of Filler Rod		Rate of Welding		Distance Between Edges		Welding Rod Used per Metre	per Foot	Edge Preparation
(mm)	(in)	(l/h)	(ft³/h)	(mm)	(in)	(m/h)	(ft/h)	(mm)	(in)	(m)	(ft)	
0·8	$\frac{1}{32}$	28–57	1–2	1·6	$\frac{1}{16}$	6–7·6	20–25	—	—	0·3	1	square
1·6	$\frac{1}{16}$	57–86	2–3	1·6	$\frac{1}{16}$	7·6–9	25–30	1·6	$\frac{1}{16}$	0·53	1·75	square
2·4	$\frac{3}{32}$	86–140	3–5	1·6	$\frac{1}{16}$	6–7·6	20–25	2·4	$\frac{3}{32}$	8·84	2·75	square
3·2	$\frac{1}{8}$	140–200	5–7	3·2	$\frac{1}{8}$	5·4–6	18–20	3·2	$\frac{1}{8}$	0·50	1·65	square
4·0	$\frac{5}{32}$	200–280	7–10	3·2	$\frac{1}{8}$	4·6–5·4	15–18	3·2	$\frac{1}{8}$	0·64	2·1	80° vee
5·0	$\frac{3}{16}$	280–370	10–13	3·2	$\frac{1}{8}$	3·6–4·6	12–15	3·2	$\frac{1}{8}$	1·40	4·8	80° vee

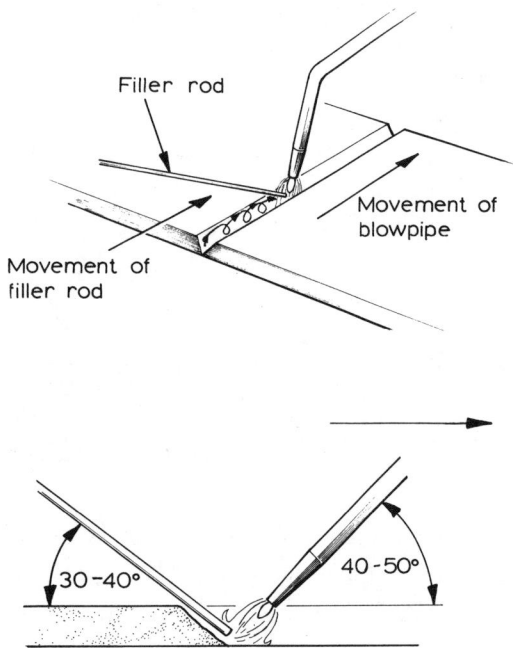

Rightward Welding

In this method, the welding is commenced on the *left-hand* side of the seam, working towards the right-hand side, as shown in the top illustration. The blowpipe points in the direction of the welded seam and moves in a straight line along the seam. The filler rod is held at an angle of 30–40° (see second illustration) and describes a series of loops as it is moved forward.

When using this method it is not necessary to bevel the edges of the plates for thicknesses up to $\frac{5}{16}$ in, and for plates over this size a vee with an included angle of only 60° is required.

The differences in edge preparation, blowpipe direction and filler rod movement are the major factors to be considered when comparing the rightward and leftward techniques.

It is more common to use the rightward technique, in preference to the leftward, on mild steel plate over $\frac{3}{16}$ in, because: (a) less filler rod is used; (b) welding speed is greater; (c) gas consumption is less; (d) the mechanical properties of the welds are better, due to the annealing effect of the flame being directed on to the completed weld; (e) there is less risk of oxidation, due to the reducing flames of the oxy-acetylene blowpipe.

Rightward Technique

Thickness of Plate		Power of Blowpipe (gas consumption)		Diameter of Filler Rod		Rate of Welding		Distance Between Edges		Welding Rod Used per Metre	per Foot	Edge Preparation
(mm)	(in)	(l/h)	(ft³/h)	(mm)	(in)	(m/h)	(ft/h)	(mm)	(in)	(m)	(ft)	
5·0	$\frac{3}{16}$	370–520	13–18	2·4	$\frac{3}{32}$	3·6–4·6	12–15	2·4	$\frac{3}{32}$	1·03	3·4	no vee
6·5	$\frac{1}{4}$	520–570	18–20	3·2	$\frac{1}{8}$	3·0–3·6	10–12	3·2	$\frac{1}{8}$	1·03	3·4	no vee
8·2	$\frac{5}{16}$	710–860	25–30	4·0	$\frac{5}{32}$	2·1–2·4	7–8	4·0	$\frac{5}{32}$	1·03	3·4	no vee
10·0	$\frac{3}{8}$	1 000–1 300	35–45	5·0	$\frac{3}{16}$	1·8–2·1	6–7	3·2	$\frac{1}{8}$	1·22	4·0	60° vee
13·0	$\frac{1}{2}$	1 300–1 400	45–50	6·5	$\frac{1}{4}$	1·3–1·5	4·5–5	3·2	$\frac{1}{8}$	1·69	4·75	60° vee
16·2	$\frac{5}{8}$	1 600–1 700	55–60	6·5	$\frac{1}{4}$	1·1–1·3	3·75–4·25	3·2	$\frac{1}{8}$	2·05	6·75	60° vee
19	$\frac{3}{4}$	1 700–2 000	60–70	6·5	$\frac{1}{4}$	0·9–1·0	3–3·25	3·2	$\frac{1}{8}$	2·96	9·75	60° top, 80° bottom vee
25	1	2 000–2 500	70–90	6·5	$\frac{1}{4}$	0·6–0·7	2–2·25	3·2	$\frac{1}{8}$	5·08	16·75	60° top, 80° bottom vee

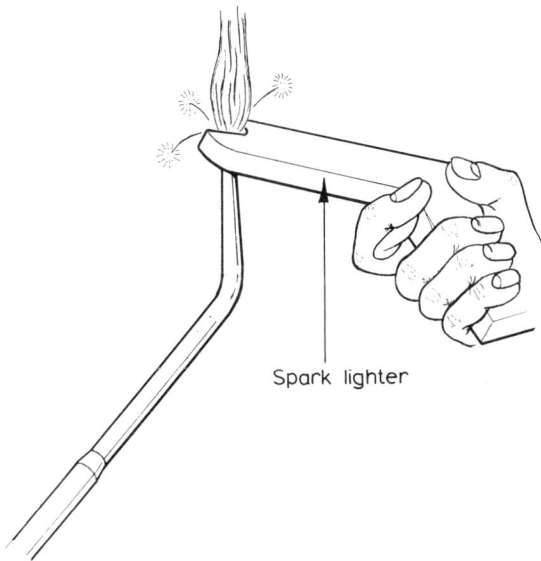

Spark lighter

Lighting and Adjusting the Flame

1   Open the blowpipe valves and turn on the gas supplies.
2   Adjust the regulators until the correct pressures are obtained. When these pressures have been set, the blowpipe valves are closed.
3   Open the acetylene valve on the blowtorch about three-quarters of a turn.
4   Allow the acetylene to flush the system through and then, using a spark lighter (or similar)—see top illustration—light the gas.
5   Open the oxygen control valve and adjust until the required flame is achieved (see page 91).

Extinguishing the Flame

If the welding equipment is to be turned off for short periods the following procedure should be observed:

(a)   Close the acetylene control valve.
(b)   Close the oxygen control valve.

Alternatively, the blowpipe may be hung on a gas economizer valve control lever if fitted.

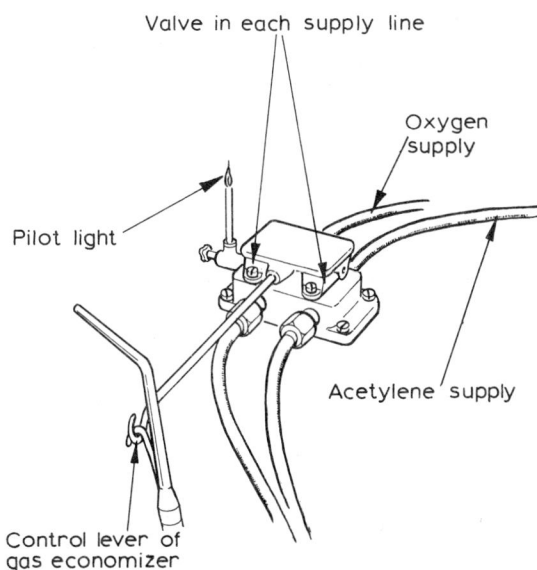

Valve in each supply line

Oxygen supply

Pilot light

Acetylene supply

Control lever of gas economizer

Closing Down Procedure

1   Repeat the procedure as in (a) and (b) above.
2   Close the cylinder valves.
3   Release the pressure in the hose by opening the control valve on the oxygen line and then closing it. Then repeat this with the acetylene control valve.
4   Release the pressure on the regulator diaphragm by turning the regulator to the minimum pressure position.

Nozzle Size, Working Pressure and Gas Consumption for Various Thicknesses of Mild Steel Plate

Thickness (mm)	(in)	Nozzle Size	Regulator Pressure (bar)	(lbf/in²)	Gas Consumption (l/h)	(ft³/h)
0·8	$\frac{1}{32}$	1	0·14	2	29	1
1·2	$\frac{3}{64}$	2	0·14	2	57	2
1·6	$\frac{1}{16}$	3	0·14	2	86	3
2·4	$\frac{3}{32}$	5	0·14	2	140	5
3·2	$\frac{1}{8}$	7	0·14	2	200	7
4·0	$\frac{5}{32}$	10	0·21	3	280	10
5·0	$\frac{3}{16}$	13	0·28	4	370	13
6·5	$\frac{1}{4}$	18	0·28	4	520	18
8·2	$\frac{5}{16}$	25	0·42	6	710	25
10·0	$\frac{3}{8}$	35	0·63	9	1 000	35
13	$\frac{1}{2}$	45	0·35	5	1 300	45
19	$\frac{3}{4}$	55	0·43	6	1 600	55
25	1	70	0·49	7	2 000	70
25+	1+	90	0·63	9	2 500	90

Note: These figures are an approximate guide and may be varied according to the flame setting, material, etc.

Nozzle Size and Gas Velocity

The power of the blowpipe is measured by the amount of acetylene gas that is used per hour. Blowpipe manufacturers have various ways of showing this:

(a) By the quantity of gas passed per hour, in litres.
(b) By the diameter of the hole in the nozzle.
(c) By a number corresponding to the plate thickness that can be welded by that nozzle.

It is most important when making a weld that the correct blowpipe power is used. The two factors that mostly affect this are the velocity and quantity of gas which passes through the nozzle.

The velocity of the gases leaving the nozzle can vary between 60 and 200 m/sec. If the lower velocity is used we get what is called a soft flame which burns quietly. Too low a pressure can give rise to backfiring in the blowpipe.

If the higher velocity is used a hard flame results which burns with a harsh hissing sound. It has a tendency to scatter the molten metal instead of allowing it to flow.

The best results are usually obtained using a flame of average velocity.

The Oxy-Acetylene Flame

It is most important when welding that the correct flame is selected to complete the welding process. Following are the three types of flame used by the welder:

Neutral Flame

This type of flame is produced when equal volumes of oxygen and acetylene are burnt together. In this flame there are two distinct zones: the inner white cone, which is clearly defined, and the outer envelope.

The neutral flame is used when welding steels, stainless steel, cast iron, copper and aluminium, etc.

Note white excess acetylene flame surrounding cone

Carburizing Flame

If the volume of oxygen being supplied to the neutral flame is reduced, the resulting flame is rich in acetylene. It is rich in carbon and capable of yielding carbon to the steel being welded. Its effect is to reduce the melting temperature of the steel in the localized area of the weld. The carburizing flame is recognized by the ragged bluish white feather surrounding the large central white cone.

The carburizing flame is used for hard facing.

Note short pointed cone and smaller flame

Oxidizing Flame

If the neutral flame is adjusted until it burns with an excess of oxygen, the flame produced is called an oxidizing flame. It can be recognized by the small white cone surrounded by the ragged bluish white feather. It tends to be hotter than the neutral flame.

The oxidizing flame is used to weld brass.

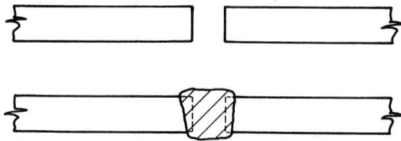

Butt Joints

Square Butt Joints

Used for plates up to 8·2 mm ($\frac{5}{16}$ in) thickness.
Plates are spaced approximately 3·2 mm
($\frac{1}{8}$ in) apart before welding, and care must be
taken to ensure that fusion is obtained
through to the underside of the plate.

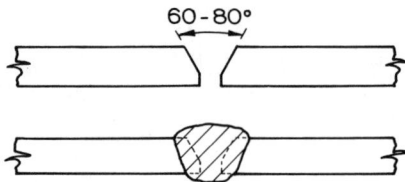

Single-vee Butt Weld

Used for plates up to 15·8 mm ($\frac{5}{8}$ in) thick.
The angle of the vee depends upon the
technique being used, the plates being spaced
approximately 3·2 mm ($\frac{1}{8}$ in) apart.

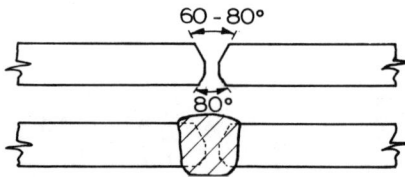

Double-vee Butt Weld

Used for plates over 13 mm ($\frac{1}{2}$ in) thick when
the welding can be performed on both sides of
the plate. The top vee angle is either 60° or
80°, while the bottom angle is 80°, depending
on the technique being used.

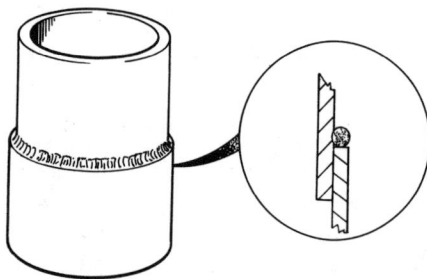

Lap Joints

Single-lap Joint

This joint, made by overlapping the edges of
the plate, is not recommended for most work.
The single lap has very little resistance to
bending. It can be used satisfactorily for
joining two cylinders that fit inside one
another.

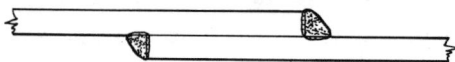

Double-lap Joint

This is stronger than the single-lap joint but
has the disadvantage that it requires twice as
much welding.

Tee Fillet Weld

This type of joint, although widely used,
should not be employed if an alternative
design is possible.

92

Chemical Changes in Metal Due to the Oxy-Acetylene Flame

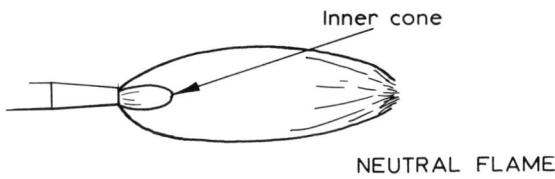

Inner cone

NEUTRAL FLAME

The Oxy-Acetylene Flame

When equal volumes of oxygen and acetylene are supplied to the mixing chamber of a blowpipe, the flame produced at the nozzle is as shown opposite, i.e. a neutral flame. The tip of the inner cone is the hottest part, having a temperature of approximately 3 300°C.

ACETYLENE+OXYGEN
= HYDROGEN+CARBON MONOXIDE

During the combustion that takes place in this part of the flame, both hydrogen and carbon monoxide are produced.

As both of these gases are combustible they continue to burn, drawing oxygen from the surrounding atmosphere.

HYDROGEN+OXYGEN
= WATER VAPOUR

The hydrogen combines with oxygen to form water vapour.

CARBON MONOXIDE+OXYGEN
= CARBON DIOXIDE

Also, the carbon monoxide combines with oxygen to form carbon dioxide.

Reducing gases

Both the hydrogen and carbon monoxide formed in the first stage of combustion act as reducing gases and protect the weld from oxidation during the welding process.

Excess of carbon

If the volume of oxygen supplied is less than the volume of acetylene, then the flame produced at the nozzle is rich in carbon (carburizing flame).
Some of the excess carbon will combine with the surface of the red-hot steel. This has the effect of lowering the melting point of the surface layer of the steel and the surface appears to froth and boil. This flame is used for special purposes, such as hard-surfacing.

Excess oxygen in reducing zone

When the reverse takes place, that is when the oxygen volume is greater than the acetylene volume (oxidizing flame), the flame burns with an excess of oxygen which is dispersed in the reducing gases zone. The free oxygen left over from the combustion process combines with the carbon and iron in the steel and forms oxides.

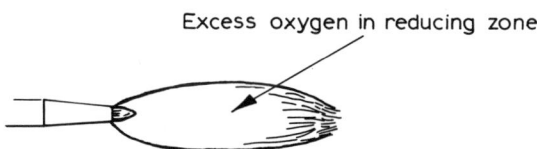

E8   Arc Welding   E8

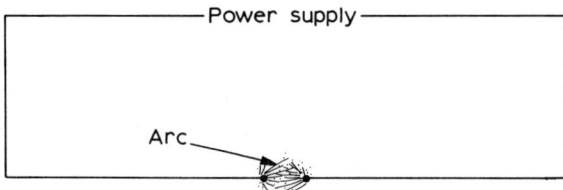

Power supply

Arc

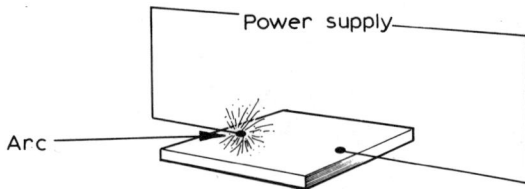

Power supply

Arc

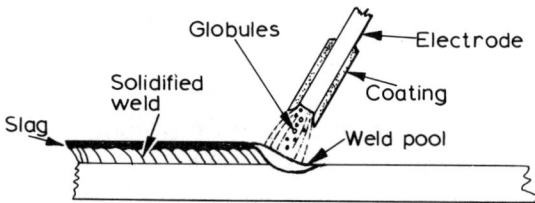

Globules

Solidified weld

Slag

Electrode

Coating

Weld pool

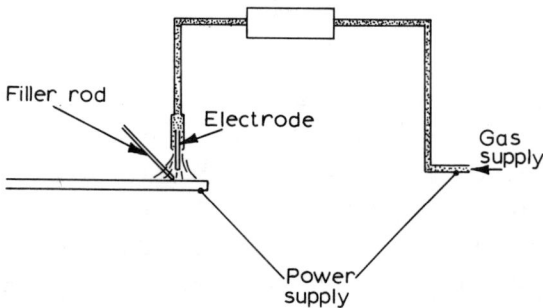

Filler rod

Electrode

Gas supply

Power supply

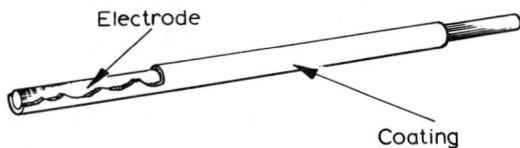

Electrode

Coating

The Manual Metal Arc Process

When two wires which form part of an electrical circuit are brought together and then pulled slowly apart, an electric spark is produced across their ends.

This spark, or arc as it is called, has a temperature of up to 3 600°C. As the arc is confined to a very small area it can melt metal almost instantly.

If one of these wires is connected to the job and the other to a wire rod or electrode, as it is usually called, the heat of the arc melts both the metal of the job and the point of the electrode. The molten metal from the electrode mixes with that from the job and forms the weld. It is important to realize that tiny globules of the molten metal from the electrode are forced through the arc (they do not fall by gravity). If this were not so it would be impossible to use this process for overhead welding.

Electrodes

When a piece of metal is heated in the atmosphere it combines with the oxygen and nitrogen to form oxides and nitrides which combine with the metal. If these were allowed to form in the weld it would result in a poor-quality, weak and brittle weld.

It is therefore necessary to protect the weld area from the air. This can be done either by surrounding the weld area by an inert gas or by the use of suitable fluxes.

It is usual, with manual metal arc welding, to use coated electrodes. These electrodes consist of a metal core surrounded by a layer of suitable flux coating.

Slag

Weld

Filler metal

Coating

Solidified weld

Weld pool

Droplets of filler metal

Slag

Arc

Protective gas

Workpiece

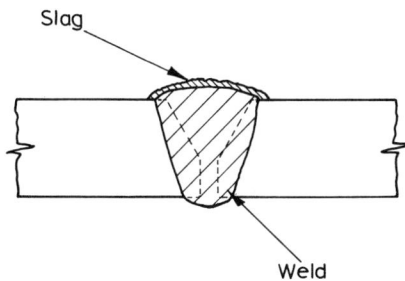

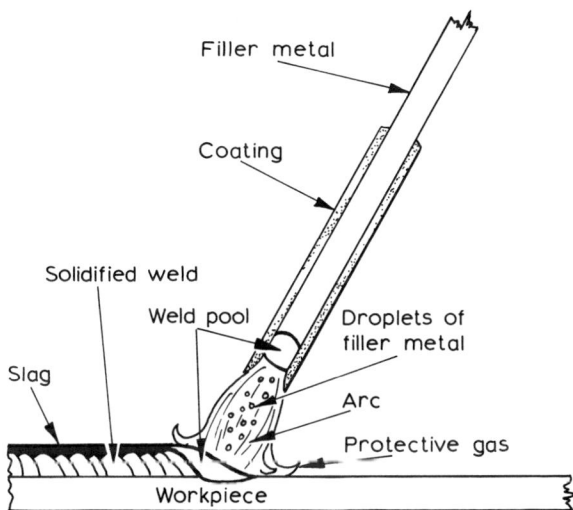

Functions of the Electrode Coating

The six main functions of the electrode coating are as follows:

1   To act as a flux and remove the impurities from the surfaces being welded.
2   To form a slag over the weld, which does the following: protects the molten metal from contact with the air; slows down the cooling rate of the weld, helping to prevent brittleness of the weld; and provides a smoother surface by preventing ripples caused during the welding process.
3   It forms a neutral gas atmosphere, which helps to protect the molten weld pool from oxygen and nitrogen in the surrounding air.
4   It helps to stabilize the arc, allowing a.c. to be used.
5   It can add certain constituents to the weld by replacing any lost during the welding process.
6   It can speed up the welding process by increasing the speed of melting of the metal and the electrode.

Manual Metal Arc Welding Equipment

To create the arc discussed earlier it is necessary to have a voltage to drive the current (which supplies the required heat) through the circuit. A voltage of between 60 and 100 V is required to create the arc, but once it has been obtained only 20–40 V are required to maintain it.

The following stages occur when creating an arc:

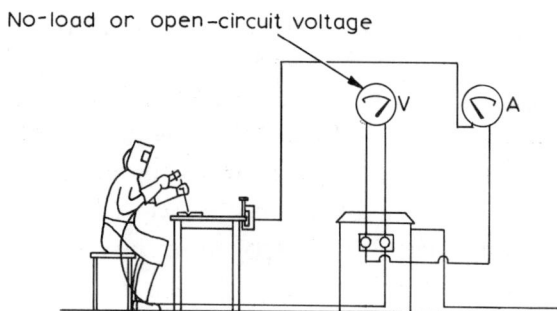

No-load or open-circuit voltage

1   With the welding plant switched on, and before welding commences, no current passes through the leads and the ammeter reads zero. A voltage has been applied to the circuit, however, and the voltmeter will read the open-circuit or no-load voltage (i.e. between 60 and 100 V).

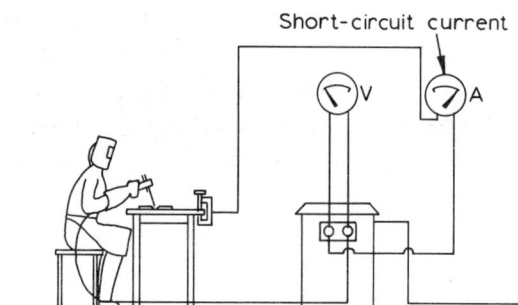

Short-circuit current

2   When the electrode is brought into contact with the job a large current, called the short-circuit current, passes through the leads, and the ammeter will deflect a large amount. While this is happening, however, the voltage drops almost to nothing. The tip of the electrode becomes hot because of the resistance created between it and the job.

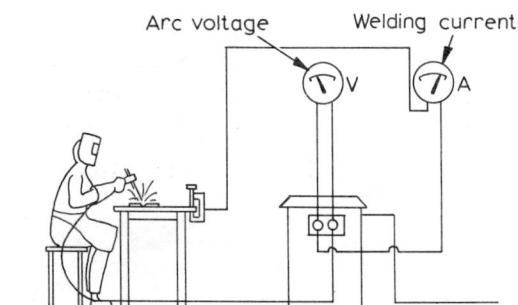

Arc voltage   Welding current

3   If the electrode is slightly withdrawn an arc is formed between the electrode and the job. The air between the two conducts the welding current. As the arc is formed the voltage rises to between 20 and 40 V and the current falls to the value to which it has been set (i.e. the welding current).

The arc is then in the normal welding condition. The heat generated by the arc melts both the workpiece and the electrode, and metal is deposited in the weld pool. During the depositing of the weld metal, variations in both the voltage and current of the arc can occur and the welding plant must be capable of coping with these changes.

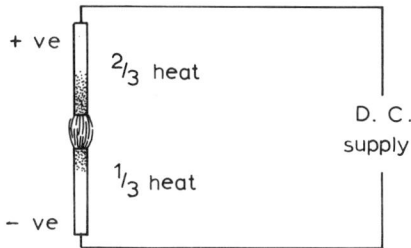

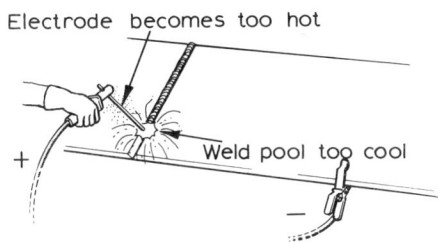

Electrode becomes too hot

Weld pool too cool

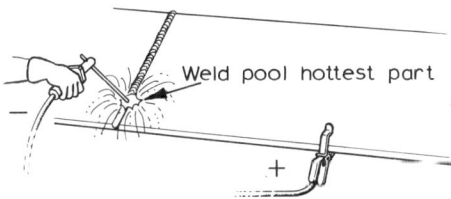

Weld pool hottest part

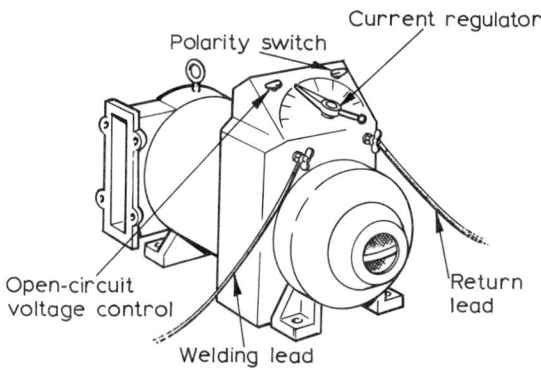

D.C. POWER SOURCE - GENERATOR

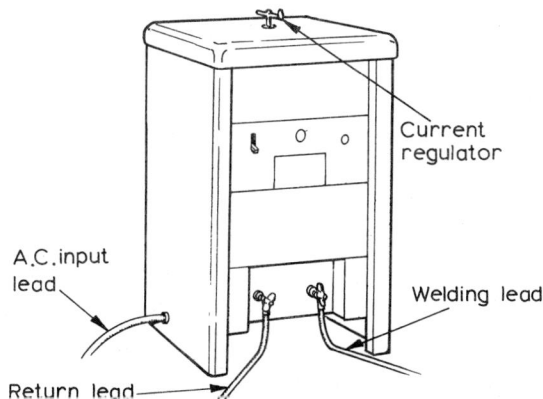

D.C. POWER SOURCE – RECTIFIER

Types of Welding Plant

(*a*)   Direct Current (d.c.)

With this system the current passes in one direction only. The heat generated is split into two parts, two-thirds goes to the positive pole and one-third to the negative pole. This is important as it determines the design of the electrode to be used.

If a light-coated electrode is connected to the positive terminal it quickly becomes too hot to use for welding. But if the workpiece is connected to the positive terminal and the electrode to the negative terminal, the weld pool becomes the hottest part and the electrode stays beneath its critical heat value.

The polarity of the electrode, when using d.c. for welding, is most important and the electrode manufacturer's recommendations should be strictly adhered to, except in exceptional circumstances where the work must be kept as cool as possible. The terms used by British Standards for electrodes state that:

Electrodes connected to the positive terminal are called electrode positive, and electrodes connected to the negative terminal are called electrode negative.

Basic equipment—1   a generator driven either from d.c. mains (motor generator) or by a petrol or diesel engine; 2   an a.c./d.c. motor generator set; or 3   rectifying equipment.

The generator must supply an open-circuit voltage of about 60 V, which will drop to approximately 20 V when the arc is struck. Generators can be obtained with various current ratings from 100 to 600 A, and the modern types automatically adjust themselves to allow for the voltage fluctuations of the arc. Normally only one welder can work from a set.

A.C. supply

Primary winding
Secondary winding
Regulator
Welding lead
Return lead

Current regulator, fine
Current regulator, coarse
Open-circuit voltage switch
Primary cable from mains supply
Return lead
Welding lead

A.C. POWER SOURCE

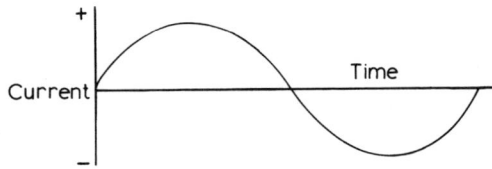

(*b*)   Alternating Current (a.c.)

With a.c., the direction of the current flow continually changes. This reverse in direction takes place 50 times per second. Because of this reversal of the current flow the two poles are maintained at the same temperature, and reversal of the terminals has no effect as is the case with d.c.

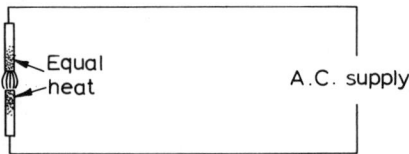

The a.c. plant consists of a transformer which will reduce the supply voltage down to the required open-circuit voltage, i.e. 60–100 V. Various types of set are available giving different current ranges. Depending on the set, current values from 20 A up to 500 A can be obtained.

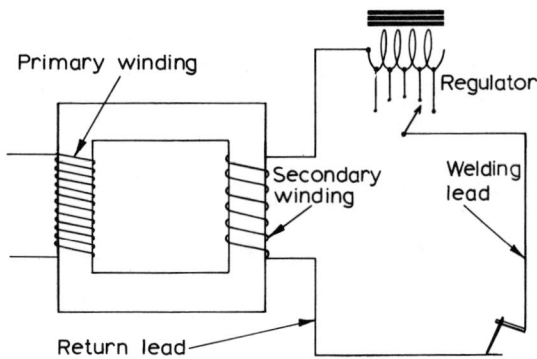

Alternating-current welding plant is cheaper to buy than the equivalent direct-current set, requires less maintenance, is quieter in operation, and the running costs are lower. The use of a.c. equipment is dependent upon an a.c. supply being available, and therefore when welding on sites it is not usually possible and a d.c. engine-driven generator is used.

Effect of Short Circuiting

When the welder strikes his arc the welding generator is subjected to a short circuit and the current passing through the windings of the generator increases. If this increase in current is not controlled the windings will overheat, resulting in damage to the generator. In most cases this short-circuit current should not exceed 150% of the normal welding current and overload devices are fitted to protect the equipment.

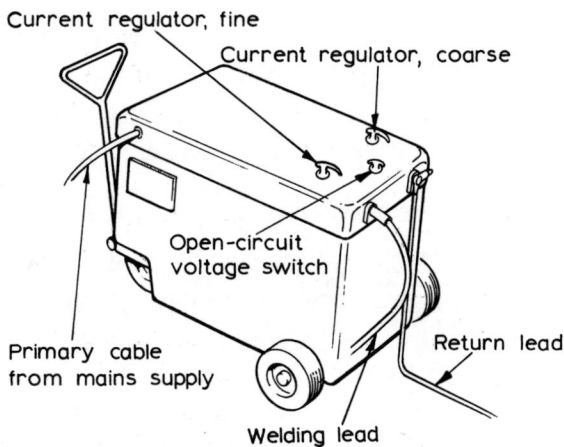

Effect of Variable Factors in Manual Metal Arc Welding

Section

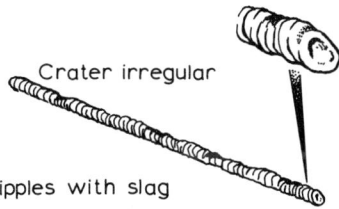

Crater irregular

Irregular ripples with slag
trapped in valleys

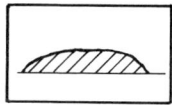

Section

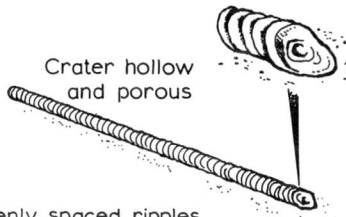

Crater hollow
and porous

Coarse, evenly spaced ripples

Section

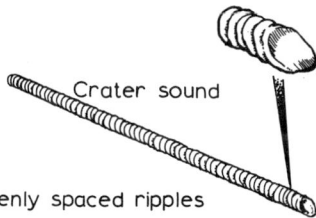

Crater sound

Smooth, evenly spaced ripples

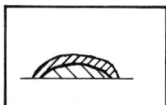

Section

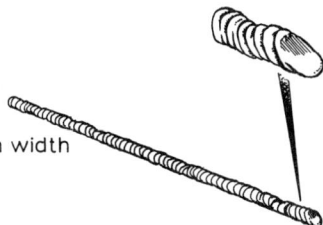

Ripples irregular in width
and in height

ARC TOO SHORT

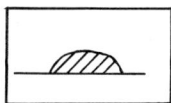

Section

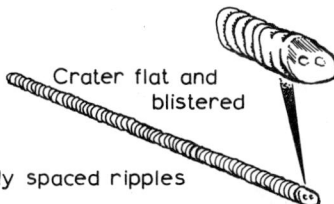

Crater flat and
blistered

Coarse, evenly spaced ripples

ARC TOO LONG

Current too Low

If the current value is too low the resulting weld has poor penetration, due to the lack of heating to create complete fusion. The weld filler metal tends to heap up on the surface of the plate without fusing to it and the arc has an unsteady sputtering sound.

Current too High

When the current value used is too high the electrode becomes red hot and a large amount of spatter takes place. This can result in blowholes being formed in the plate, excessive penetration resulting in weld metal beads on the underside of the plate, undercut along the edge of the weld and excessive oxidation and slag which is hard to remove. The arc has a fierce crackling sound

Correct Current

With the correct current the arc has a steady crackling sound. The weld formed has good penetration and is easily controlled.

Arc Length

The arc length is the distance between the tip of the electrode and the surface of the weld pool. It should be approximately equal to the diameter of the wire core of the electrode being used. When this distance is correct the electrode metal is deposited in a steady stream of metal particles into the weld pool. If the arc length is reduced it becomes difficult to maintain the arc, due to the increase in welding current that takes place, and it can result in the electrode becoming welded to the weld pool. Also, if the arc length is increased the welding current is reduced, resulting in a poor weld being produced, and the protective gas shield produced from the electrode surrounding the weld pool cannot efficiently prohibit the formation of oxides, etc., in the weld.

Section

Crater porous

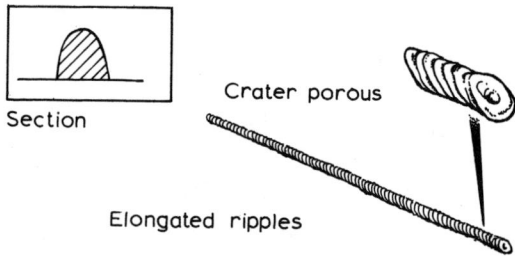

Elongated ripples

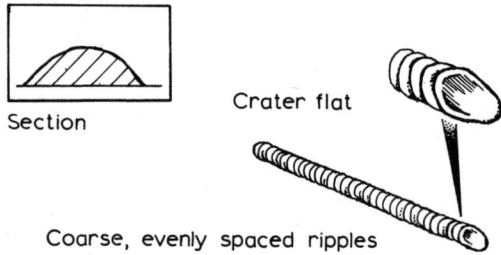

Section

Crater flat

Coarse, evenly spaced ripples

Speed of Travel

A fast rate of travel results in a thin deposit of the filler metal and can result in insufficient fusion of the filler metal with the base metal. The surface of the weld has elongated ripples and a porous crater.

Too slow a rate of travel gives a wide thick deposit of the filler and it can allow the slag to flood the weld pool making it difficult to deposit the filler metal. The surface of the weld appears as coarse ripples and has a flat crater.

Angle of Electrode

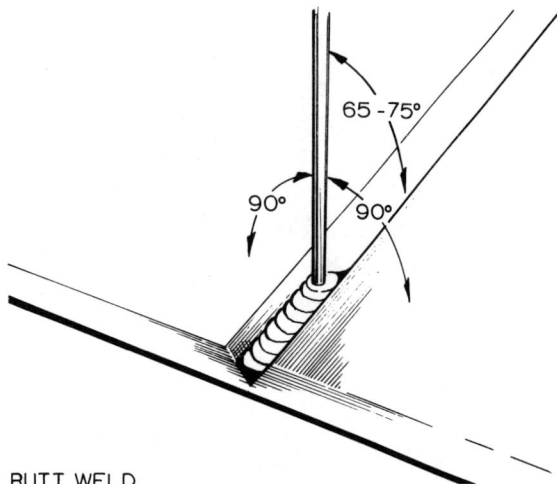

BUTT WELD

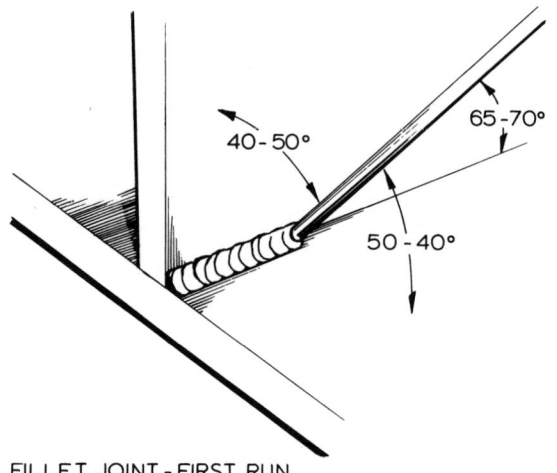

FILLET JOINT - FIRST RUN

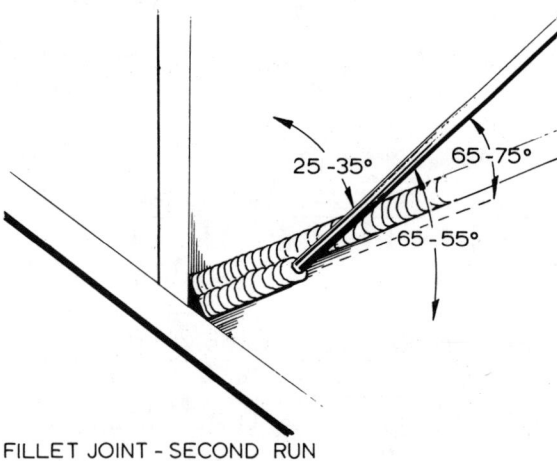

FILLET JOINT - SECOND RUN

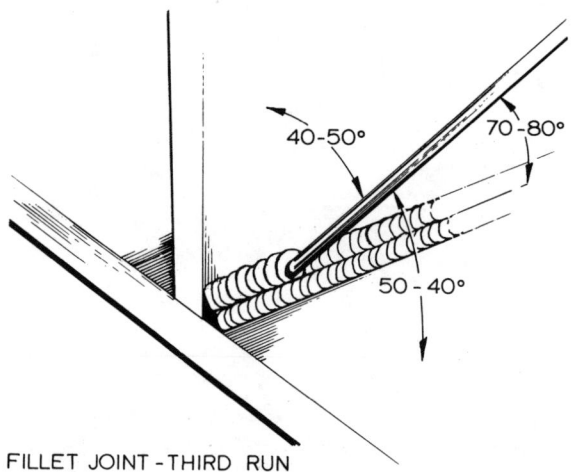

FILLET JOINT - THIRD RUN

When the weld is completed with the correct current, arc length, speed of travel and electrode angle, the deposit will be of regular thickness and width and the surface will have even, smooth ripples, and be free from slag entrapment and porosity.

E9   Welding Safety, Testing and Inspection   E9

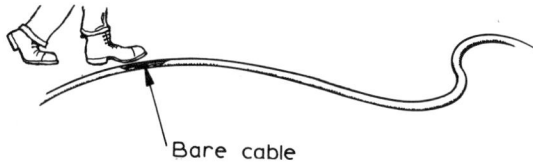

Bare cable

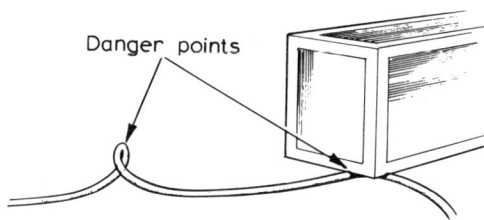

Danger points

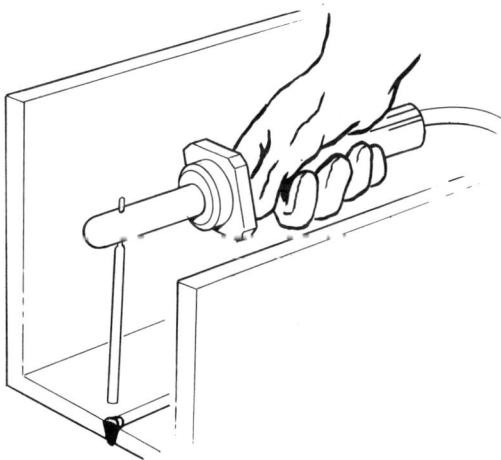

Safety in Manual Metal Arc Welding

Equipment

The power supply for arc welding may be either a.c. or d.c. The voltages used are in the order of up to 100 V and currents of up to 600 A. It is therefore essential that adequate precautions are observed to protect both the operator and other personnel in the area from the risks associated with electricity. The following points should be observed when using arc welding equipment:

1   Make sure that the welding circuit is correctly earthed.
2   The welding cables should be adequately insulated. **Never** use damaged cables. (See top illustration.)
3   Ensure that the welding cables are free from kinks and that they do not become crushed or pinched under heavy loads (second illustration).
4   Check that the power source is disconnected from the supply before attaching the cables to the terminals.
5   Always use a fully insulated electrode holder when welding in confined spaces, or where it is difficult to get access to the weld without touching the surrounding metal (third illustration).

Eye Protection

It is essential to protect the eyes and face from the arc. The use of a face mask or head shield with the correct filter glass is essential.

Protect your eyes from the arc

Recommended Filters

Shade (B.S. 679)	Metal Arc Current
9 EW	up to 100 A
11 EW	100–300 A
13 EW	over 300 A

Screening and Protective Clothing

Adequate screening around the welding area should be erected to protect the workers in the immediate vicinity. The operator should also be protected from the splashing metal and sparks given off from the welding by wearing a leather apron, spats and gloves. When working on a site or in damp conditions the use of rubber-soled boots is also advisable.

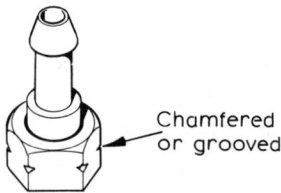

LEFT-HAND THREAD
COMBUSTIBLE GAS

Chamfered
or grooved

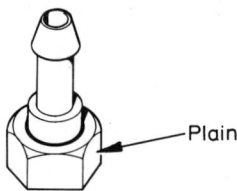

RIGHT-HAND THREAD
NON-COMBUSTIBLE GAS

Plain

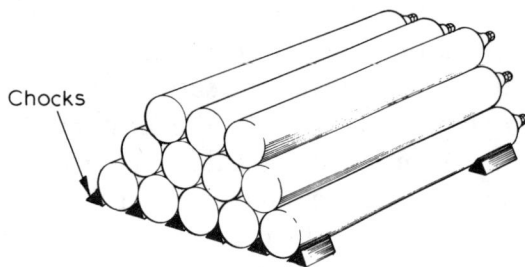

Chocks

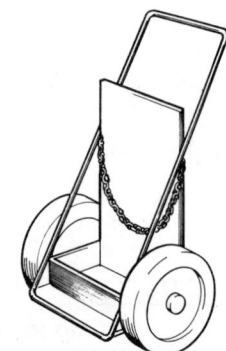

PORTABLE
WELDING TROLLEY

Care of Gas Cylinders

Identification, Handling and Storage

Cylinders containing gases are manufactured under strict government regulations. A colour code (B.S.349: 1932) is used to indicate the contents of each cylinder.

Black	*Oxygen*
Maroon	*Acetylene*
Red	*Coal gas and Hydrogen*
Dark Grey with	
Black neck	*Nitrogen*
Grey	*Air*

To prevent the interchange of fittings between cylinders containing combustible and non-combustible gases, the valve outlets on combustible gas cylinders are screwed left-handed and on non-combustible gas cylinders right-handed.

Do not

alter the colour of a cylinder.

Do not

tamper with the valve threads of a gas cylinder.

Oxygen and acetylene cylinders should never be stored together. They should be stacked as in the third diagram, with the bottom layer well chocked to prevent rolling, or stored upright with a securing chain to prevent them falling over. The full cylinders should be stored separately from empty ones and used up in rotation as received. The storage area should be cool and dry, protected from direct sunlight, frost, rain and corrosive conditions. There should be prominent signs indicating the presence of a fire risk and *No Smoking* signs should be strictly observed.

Cylinders should never be handled roughly or allowed to fall from a height. When using cranes, etc., to lift cylinders, a protective mat or cradle sheuld be used. Cylinders should *never* be used as rollers. If a cylinder has been damaged in any way it should be labelled accordingly, and the suppliers should be notified when it is collected. When moving cylinders that are in use and have regulator valves attached use a proper trolley.

The cylinder valves should be turned off, using the correct key, before the cylinders are moved, and care must be taken to protect the regulators from knocks and bangs.

Fire and Explosive Risks

Oxygen

Oxygen cylinders and the fittings attached to them must *never* be greased or oiled. Even small quantities of oil and grease, when mixed with oxygen, can react and cause an explosion. Oxygen has no smell and does not itself burn, but it will support and speed up combustion. It should never be used for ventilating enclosed spaces or for blowing clothes or other combustible material (top illustration). Even a spark can, under these conditions, start a fire which may result in considerable damage or injury to personnel.

Acetylene

Acetylene gas has a noticeable smell, is highly inflammable and will ignite and burn very easily. Even a mixture of acetylene and air as low as 2% will give an explosive compound condition when in contact with metals and alloys, such as those containing copper and silver. Pipings and fittings made of copper should never be used with acetylene gas.

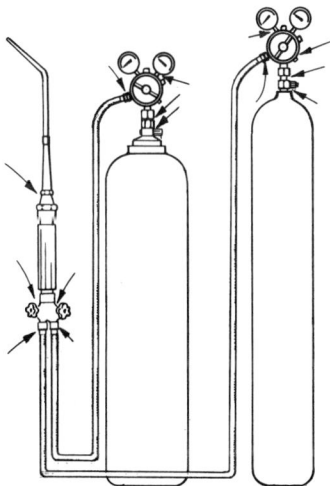

Take extra care with both oxygen and acetylene cylinders to ensure that there are no leakages when assembling the equipment. Dirt should be blown clear of the valve seats by cracking the valves momentarily before attaching the pressure regulator.

Never

test for leakage with a flame. Soapy water should be brushed over the joint to test for leakage of gas (second illustration). Always check, on completion of the job, that no glowing fragments or sparks are left.

Remember

that even at some distance away from the oxy-acetylene flame the temperature is still very high and capable of starting a fire (bottom illustration).

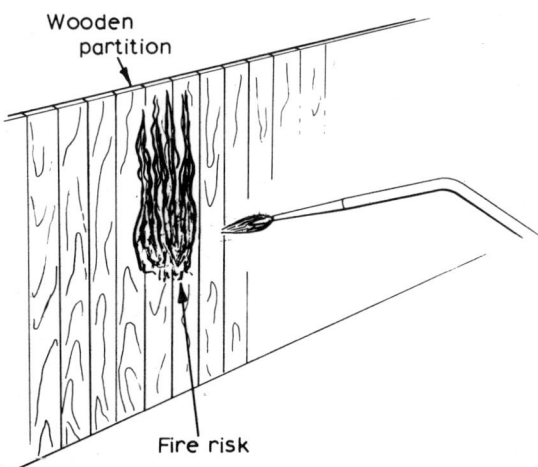

NEVER USE OXYGEN TO BLOW CLOTHING CLEAN

TEST FOR LEAKS WITH SOAPY WATER NOT WITH A FLAME

Wooden partition

Fire risk

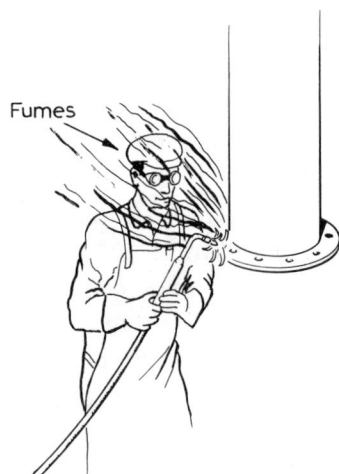

Dangers of Welding in Confined Spaces

When welding, fumes are given off which are dangerous to the health of the welder. It is therefore important to remove and dispose of these fumes. This is particularly important when welding in confined spaces, such as inside boilers, tanks, ships' compartments, etc.

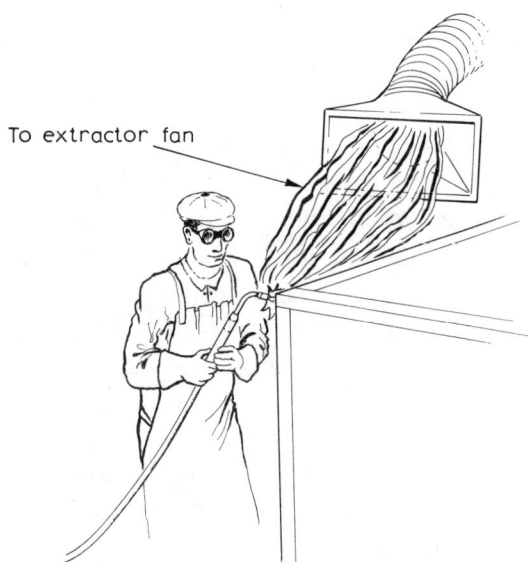

When welding in confined spaces it is necessary to use some form of mechanical ventilation. This usually consists of an elongated hood attached to a flexible ducting, connected to an extraction fan. It is necessary to arrange the hood in such a position that it draws air across the face of the welder towards the welding. Forced ventilation not only reduces the fumes but also lowers the temperature in which the welder is working and increases his comfort.

The fumes from welding consist of iron oxide particles which, when inhaled, will remain in the lungs causing damage to them. Arcing can also result in nitrous fumes (oxides of nitrogen) being formed. These give little warning of their presence and can cause serious damage to the lungs.

Oxygen should *never* be used to ventilate confined spaces. Fatal accidents through workers' clothes becoming ignited have resulted from this practice.

The use of compressed air for forced ventilation is also an undesirable practice.

NEVER USE OXYGEN TO VENTILATE CONFINED SPACES

Weld Defects

Lack of Penetration

Lack of penetration is the failure of the filler metal to penetrate into the joint. It is caused by:

Incorrect edge penetration.
Incorrect welding technique.
Inadequate de-slagging.

Lack of Fusion

Lack of fusion is the failure of the filler metal to fuse with the parent metal. It is caused by:

Insufficient heat.
Too fast a travel.
Incorrect welding technique.

Porosity

Porosity is a group of small holes throughout the weld metal. It is caused by the trapping of gas during the welding process, due to chemicals in the metal, dampness, or too rapid cooling of the weld.

Slag Inclusion

Slag inclusion is the entrapment of slag or other impurities in the weld. It is caused by the slag from previous runs not being cleaned away, or insufficient cleaning and preparation of the base metal before welding commences.

Undercut

Undercuts are grooves or slots along the edges of the weld caused by:

Too fast a travel.
Too great a heat build-up.
Bad welding technique.

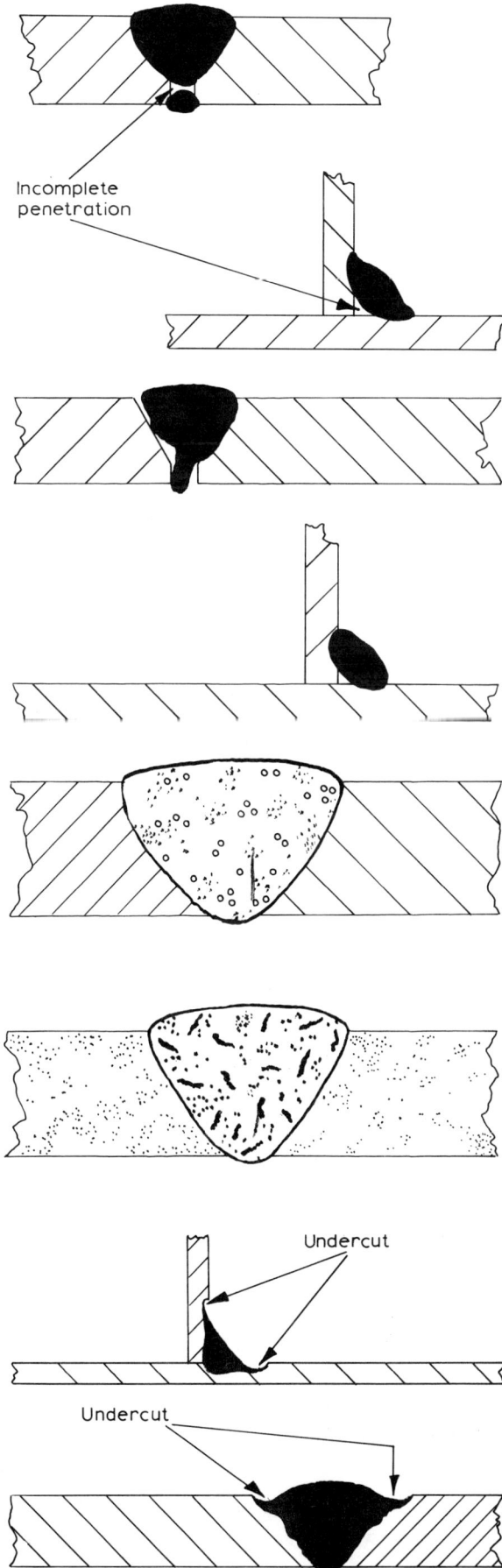

105

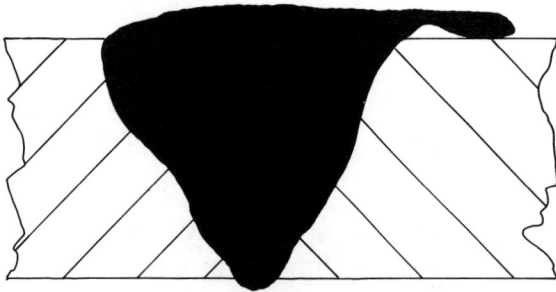

Overlays

Overlays consist of metal that has flowed on to the parent metal without fusing with it. The defect is caused by:

> Insufficient heat.
> Contamination of the surface of the parent metal.
> Bad welding technique.

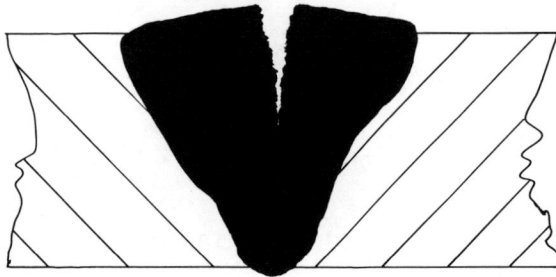

Cracking

Cracking is the formation of cracks either in the weld metal or the parent metal. It is caused by:

> Bad welding technique.
> Unsuitable parent metals used in the weld.

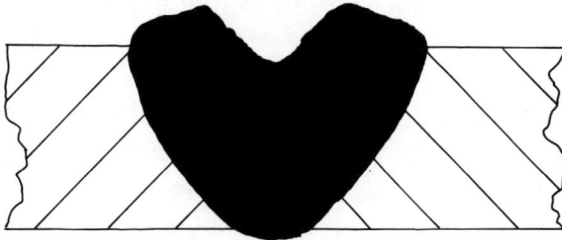

Blowholes

Blowholes are large holes in the weld caused by:

> Gas being trapped, due to moisture.
> Contamination of either the filler or parent metals.

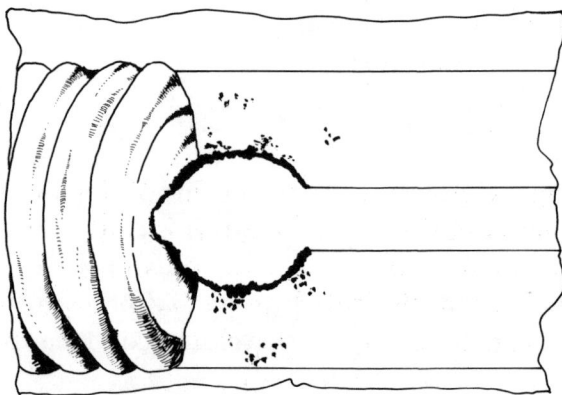

Burn Through

Burn through is the collapse of the weld pool due to:

> Poor edge preparation.
> Too great a heat concentration.

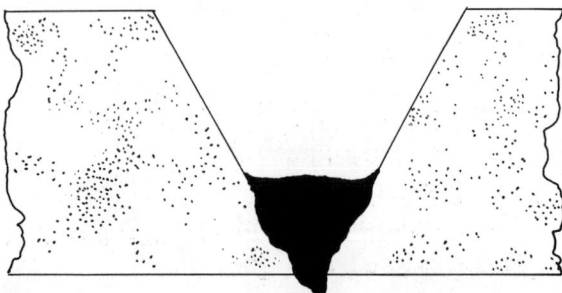

Excessive Penetration

Excessive penetration is where the weld metal protrudes through the root of the weld. It is caused by:

> Too big a heat concentration.
> Too slow a travel.
> Incorrect edge preparation.

180° BEND   110° BEND

CASUAL BENDS

BEND IN PLATE ONLY   BEND MAINLY IN WELD

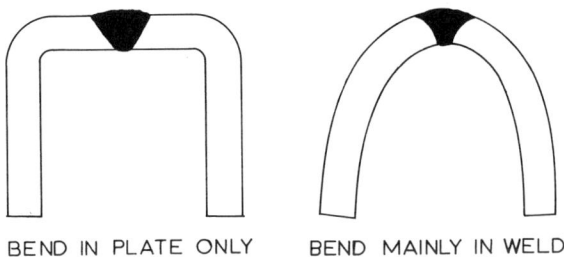

OFF-CENTRE BEND

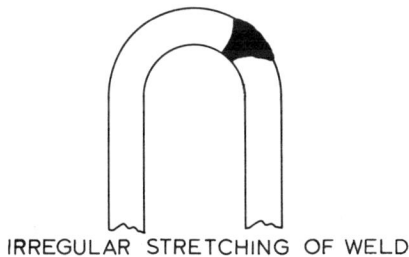

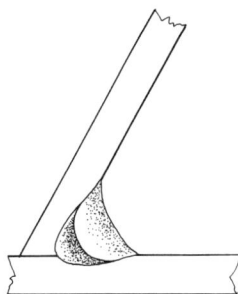

IRREGULAR STRETCHING OF WELD

Workshop Testing of Welds

Bend Test

Ideally, the test-piece should be machined on both sides and polished in the direction of its length. The polishing should be done carefully and all machine marks removed, because the slightest scratch or tool mark could result in a crack affecting the results of the test.

It is usual to bend the test-piece over a former of diameter twice the thickness of the plate, through an angle of 180°. Bending the test-piece through an angle of 180°, i.e. until both ends are practically parallel to each other, does away with the need of measuring the angle, because, after approximately 110°, only the parent metal stretches. If lower angles are used (e.g. 90–120°) it will require accurate measurement of the angle, which can be difficult.

Casual bending of the test piece in a vice can lead to either excessive or insufficient bending of the weld, as no control can be made over the radius of the bend. Uniformity of test made in this way is impossible to achieve.

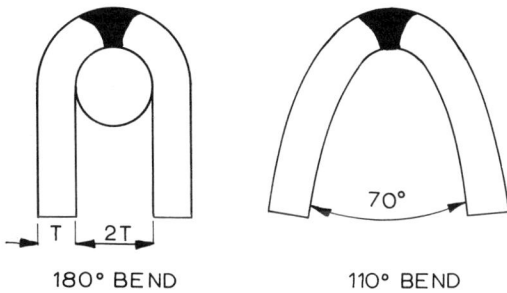

ETCHED SPECIMEN
SHOWS
1 Uniformity of fusion
2 Two runs used

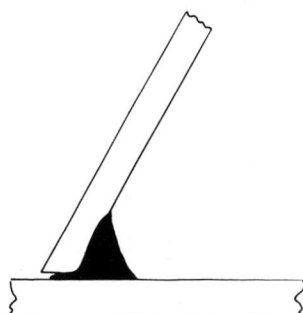

ETCHED SPECIMEN
SHOWS
1 Unfused metal between faces
2 Poor fitting of parts
3 Uneven fusion
4 Only one run used

Macroetch Examination

To prepare a specimen for examination it is necessary to initially file off any deep scratches, etc., until a smooth surface is obtained. It is then polished with successively finer grades of emery paper until a highly polished scratch-free surface is obtained. The surface is then etched in a suitable solution (e.g. 10–15 ml nitric acid to 90 ml alcohol), either by immersing in the etching fluid or by swabbing it on to the surface. It is then washed in hot water, given a rinsing in alcohol, and dried in a warm air current. The etched specimen can then be inspected with magnifying glass or naked eye, and any defect in the weld observed.

E10   Oxy-Gas Cutting

The Process

When oxygen is combined with steel, particularly if the steel has been heated, it will rapidly oxidize. The oxygen combines with the steel to form a slag or magnetic oxide. Oxygen cutting relies on this fact, and in effect the process is one of partial disintegration rather than cutting. In practice, the speed of velocity of the oxygen, together with particles of the oxide, through the kerf or cut, have a scouring effect on the sides of the cut and remove particles of partially oxidized metal adjacent to the oxygen-cut face. These factors are important and help to produce the clean cutting resulting from modern equipment. The temperature to which the steel is heated to precipitate this oxidation is approximately 900°C, which is well below the melting point of the steel and has little effect on the mechanical properties of the metal adjacent to the cut, except for a slight hardening in the surface of the cut edges.

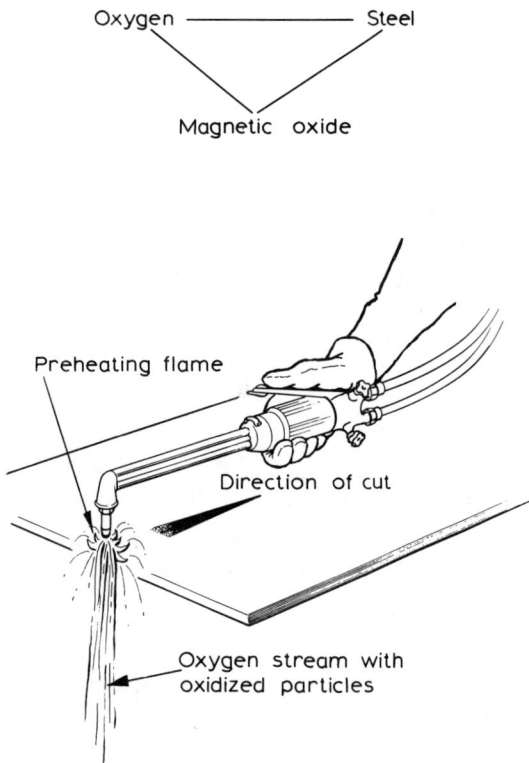

Fuel Gases

The most commonly used fuel gases are:

acetylene, propane, coal gas and hydrogen.

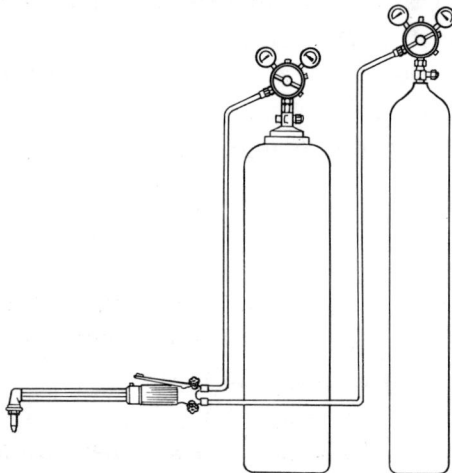

Equipment

The equipment required for oxygen cutting is similar to that described on **page 86 for** oxy-acetylene welding. It differs only in the item below.

Pressure Regulator

The oxygen regulator should be capable of supplying oxygen at a pressure of approximately 5·0 bar (70 lbf/in²) for thicknesses of up to 102 mm (4 in) and higher if it is proposed to work on thicker plate.

The acetylene regulator used for welding will be adequate for the cutting process. If coal gas is to be used from cylinders, a regulator supplying up to 2 bar (30 lbf/in²) is required.

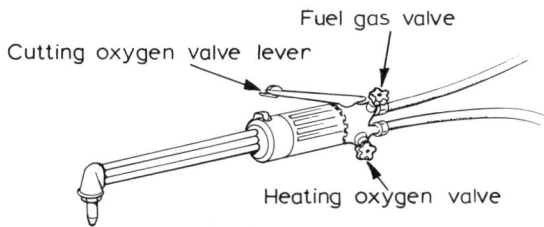

CUTTING BLOWPIPE

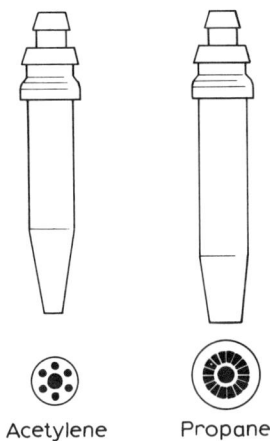

Fuel gas valve

Cutting oxygen valve lever

Heating oxygen valve

Acetylene   Propane

CUTTING NOZZLES

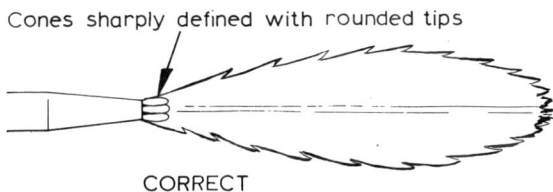

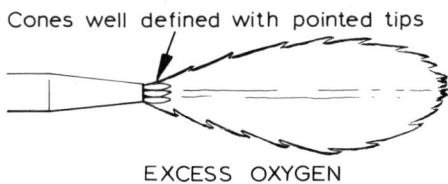

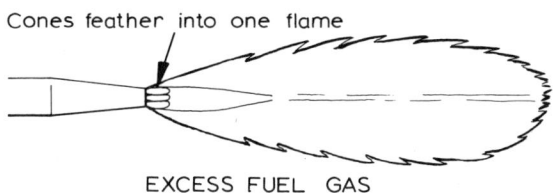

Cones sharply defined with rounded tips

CORRECT

Cones well defined with pointed tips

EXCESS OXYGEN

Cones feather into one flame

EXCESS FUEL GAS

Blowpipe

The blowpipe is of similar construction to that used for welding. The oxygen and fuel gas are mixed to form the preheating flame and a separate supply tube is incorporated to supply additional oxygen required for the cutting process. This oxygen supply is controlled by a separate valve lever on the blowpipe.

Nozzles

The nozzles used for cutting differ from the welding nozzles in that there is a central hole through which the cutting oxygen passes, and around this there are a series of holes or flutes which deliver the oxy–fuel-gas mixture for the preheating flame.

The size of the nozzles is determined by the diameter of the central hole or orifice. It is important to ensure that the correctly designed nozzle is used with the various types of fuel gas in use.

Lighting and Adjusting the Blowpipe

1 Ensure that the blowpipe valves are closed.
2 Open the cylinder valves.
3 Open the oxygen cutting valve on the blowpipe and set the oxygen working pressure on the regulator.
4 Shut the oxygen valve.
5 Set the fuel-gas regulator to the correct working pressure.
6 Open the fuel-gas valve or blowpipe and allow the system to flush through.
7 Light the gas with a spark lighter.
8 Open the heating oxygen valve and adjust until a neutral flame is achieved.
9 Open the oxygen cutting valve and adjust the fuel-gas valve to restore a neutral flame.

The cutter is then ready for use. To extinguish it turn off the cutting oxygen, close the fuel-gas valve and then close the heating oxygen valve.

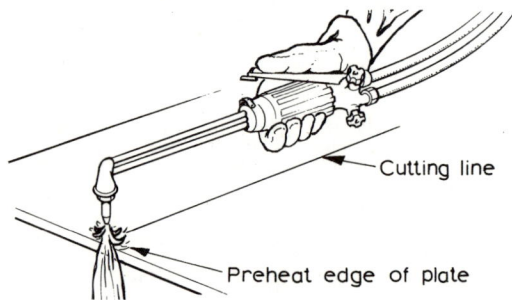

Use of Blowpipe

With the flame correctly adjusted, as indicated on the previous page, hold the blowpipe so that the tips of the cones are 3–5 mm above the edge of the plate and vertically above it.

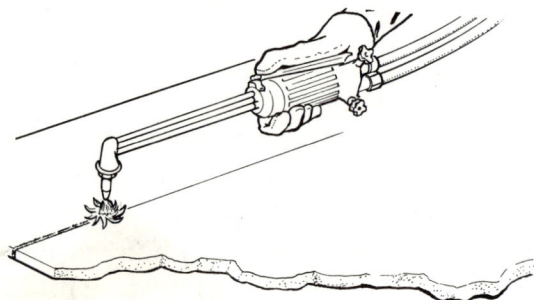

When the plate edge is heated to a dull red, turn on the cutting oxygen. Move the blowpipe at a uniform speed along the line to be cut, keeping the blowpipe nozzle at a distance of 3–5 mm from the plate.

Gouging

This process is similar to cutting except that only a groove is produced in the surface of the plate. The procedure for gouging is indicated below.

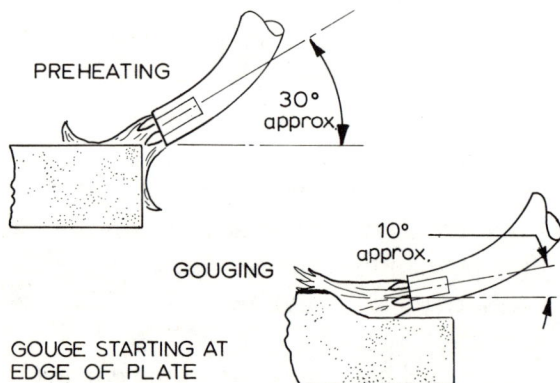

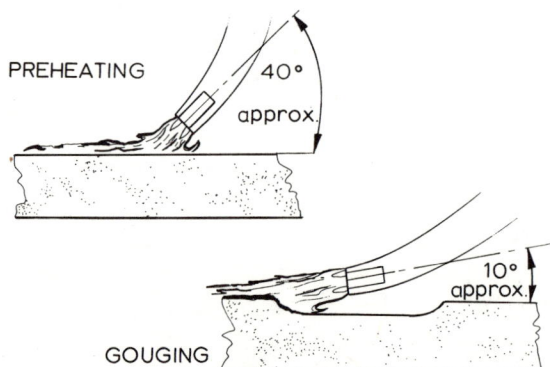

Approximate Pressures for Cutting Mild Steel Plate

| Plate Thickness | | Nozzle Size | | Gas Pressure | | | | Gas Pressure | | | |
| | | | | Acetylene | | Oxygen | | Propane | | Oxygen | |
(mm)	(in)	(mm)	(in)	(bar)	(lbf/in²)	(bar)	(lbf/in²)	(bar)	(lbf/in²)	(bar)	(lbf/in²)
6·5	$\frac{1}{4}$	0·75	$\frac{1}{32}$	0·14	2	1·8	25	0·21	3	2·5	35
13	$\frac{1}{2}$	1·0	$\frac{3}{64}$	0·14	2	2·1	30	0·21	3	2·8	40
25	1	1·5	$\frac{1}{16}$	0·14	2	2·8	40	0·28	4	3·5	50
51	2	1·5	$\frac{1}{16}$	0·14	2	3·5	50	0·28	4	4·2	60
76	3	1·5	$\frac{1}{16}$	0·14	2	4·2	60	0·35	5	5·0	70
102	4	2·0	$\frac{5}{64}$	0·28	4	5·0	70	0·49	7	5·4	75
152	6	2·5	$\frac{3}{32}$	0·28	4	5·7	80	0·49	7	6·4	90

Effects of Variation in Cutting Procedure

	APPEARANCE	CAUSE
	Sharp tapered bottom edge with square face. Smooth surface with slight drag lines. Light easily removable oxide scale.	*Correct*
	Top edge not sharp with undercutting. Bottom edge slightly rounded. Drag lines have excessive backward drag.	*Speed too fast*
	Melted and rounded lip edge. Rough bottom edge. Irregular gouging on lower half of cut face. Heavy scale, difficult to remove.	*Speed too slow*
	Excessive melting and rounding of top edge. Sharp bottom edge. Undercut at top of the cut edge.	*Nozzle too high*
	Top edge slightly rounded and beaded. Square cut face with sharp bottom corner.	*Nozzle too low*
	A regular bead along top edge. A wider kerf at the top edge with undercutting of the face.	*Cutting oxygen pressure too high*
	A rounded top edge. The melted metal running into the kerf. Smooth cut face tapered from top to bottom. Lightly adhering slag.	*Preheating flame too large*

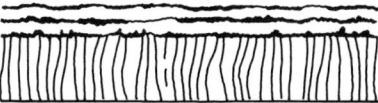

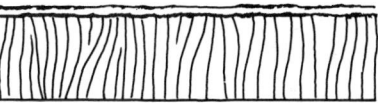

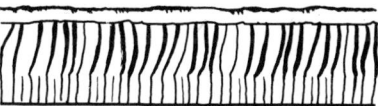

E11   Forging and Forming   E11

Tuyère

Precautions to be Observed when Forging

The first precaution in forging is to ensure that there is plenty of fuel burning between the outlet of the tuyère and the work being heated. The blast should not hit the workpiece in the same way as from a blowpipe. If it does the workpiece will oxidize and 'burn' into holes.

GRAIN SIZE OF STEEL

Oxide film

Correctly heated to forging temperature

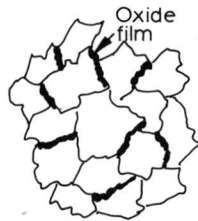

Overheated burnt metal

The temperature of the workpiece should be controlled by careful observation, and it should not be left too long in the firebed, otherwise the grains of the metal will enlarge. If grain growth occurs the forging may split or crack.

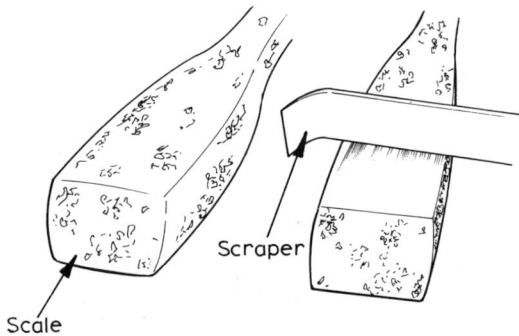

Scraper

Scale

When taken from the firebed the workpiece should first be scraped with a flat scraper to remove the oxided skin of metal (scale). If this is not done, it will, when hit with a hammer, shower red-hot sparks and pieces of metal in all directions. Some of the scale may also be forced into the metal and a flaw will be introduced. Operators of a forge should wear gloves, aprons and goggles.

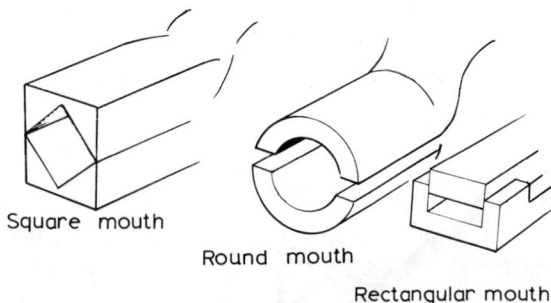

Square mouth

Round mouth

Rectangular mouth

BLACKSMITH'S TONGS

Correctly shaped tongs should be used to grip the workpiece and a locking link applied for safety.

Beware

black objects and shapes on a blacksmith's floor may be hot.

Bench Stakes

Stakes are made from metal of heavy section, and are available in a variety of shapes and sizes. Stakes are used to form sheet metal to a desired shape. A number of common stakes are available to cover the majority of shapes met with in the workshop. Work outside the range of these stakes requires the craftsman to exercise his skill in adapting suitable pieces of metal to produce the shape he requires.

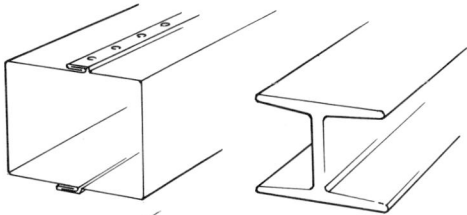

Common Stakes met with in the Workshop

1   'H' Section

A useful bench stake used for fabricating rectangular ducting, by riveting or grooving.

2   Round Section

Useful for forming circular ducting.

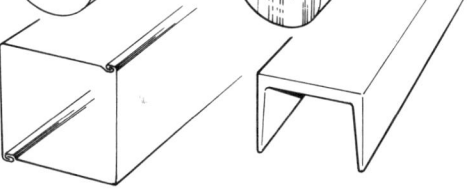

3   Channel Section

Used as the 'H' section, but having sharp corners it is useful for Knocked-up Seams.

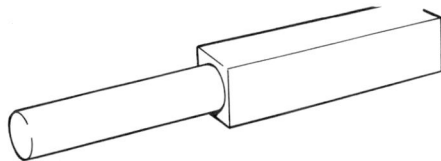

4   A Bench Bar

Manufactured with both round and rectangular sections.

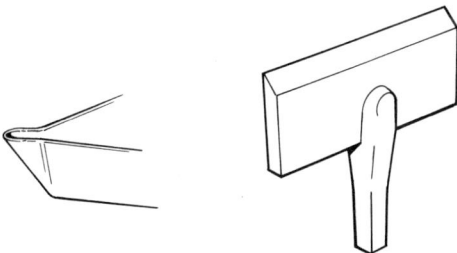

5   Hatchet Stake

To allow hand creasing or forming, e.g. pig's ear corner on a roasting tin.

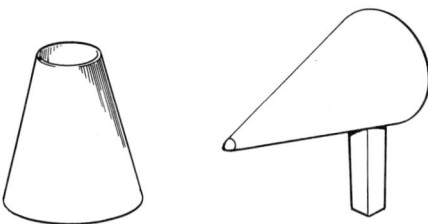

6   Funnel or Cone Stake

For hand forming of conic sections.

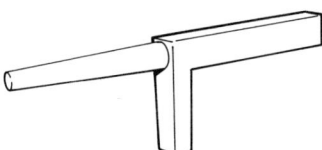

7   Beck or Bick Iron

A stake with two sections. A slow taper and rectangular.

113

Spinning and Drawing

In mass production of components, the expense of production is extremely important. Processes must become mechanized and means must be found to replace *hand-forming* processes, such as hollowing and raising from sheet metal, by machine methods. Spinning and drawing are two of these machine methods.

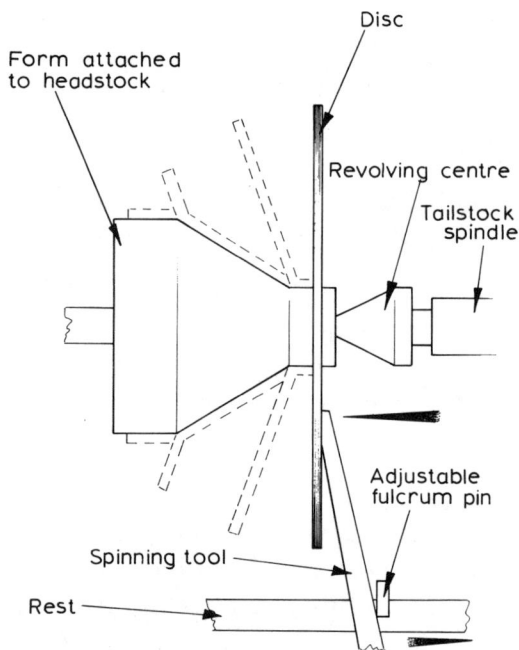

Spinning is carried out on a simple form of lathe. A circular disc or blank of metal, equal in surface area to that of the desired shape, is held by pressure from the tailstock against a form tool of the same shape as the finished article, which is attached to the faceplate (the form tool can be made of either wood or metal). The disc is rotated rapidly and pressure applied to deflect the metal on to the former. Various form tools are used, these being held against a fulcrum pin by the operator. The form is produced in several operations, the fulcrum pin being moved to different positions on the rest. Large dish-ends for vessels are produced by this method, the form tools in this case being operated either mechanically or hydraulically.

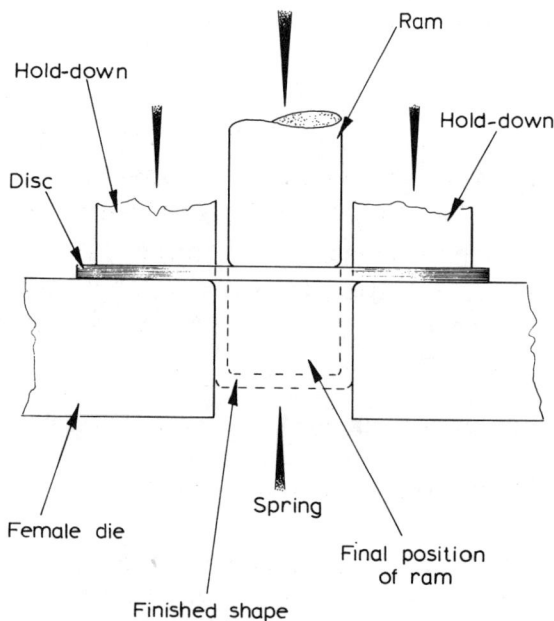

In the *drawing* process, illustrated opposite, the components are produced more quickly, although the cost of tooling is much higher than with spinning.

The disc is held against the die by a cylindrical hold-down. The ram moves into the die, pulling the disc through the hold-down.

N.B. if the hold-down pressure is too great the disc will tear, if too small wrinkling will take place.

After completing the drawing operation the ram will return to its original position. The spring or ejector mechanism will force the article out of the die. A lubricant is used in this process to assist drawing and to overcome the problems of friction.

Bending and Bend Allowance

It is important to understand that materials such as sheets and plates have thickness. This thickness must be taken into account when undertaking processes such as bending (see subsection E4, page 124, of Base book).

The present subsection gives a basic understanding of allowance required for bending. A more detailed treatment of this topic will be given at a later stage of training.

The top illustration shows a simple section formed by bending. The cutting size will be 60 mm. (N.B. The dimensions given are internal.)

The second illustration shows a similarly formed section. This time the dimensions are external. The cutting size is now 54 mm, i.e. $(30-3)+(30-3)$. The two metal thicknesses have been deducted.

The third illustration shows a more complex section. As it is drawn to internal dimensions the cutting size will be:

$$50+30+30 = 110 \text{ mm.}$$

The fourth illustration shows the same section dimensioned externally. The cutting size will now be:

$$(30-3)+(30-3)+(50-6) = 98 \text{ mm.}$$

(N.B. 110 mm—4 metal thicknesses)

When forming metal to a radius, consideration must be given to the mean line, which is the boundary line between the area under compression and the area under tension. For curves of normal radius the mean line may be taken as $0.5t$, where t is the metal thickness. In the illustration opposite, the cutting size to form this curve will be:

$$\frac{2\pi r}{4} = \frac{2\pi\,41.5}{4} = 65.2 \text{ mm.}$$

(N.B. 41.5 is the mean radius line.)

To calculate the cutting size of the component shown in the bottom illustration divide the component into parts

$$
\begin{aligned}
AB &= 2\pi r/4 = \pi 28.5/2 &= 44.8\\
BC &= 100-(30+30) &= 40.0\\
CD &= 30-6 &= 24.0\\
DE &= 30-3 &= \underline{27.0}\\
& &135.8 \text{ mm.}
\end{aligned}
$$

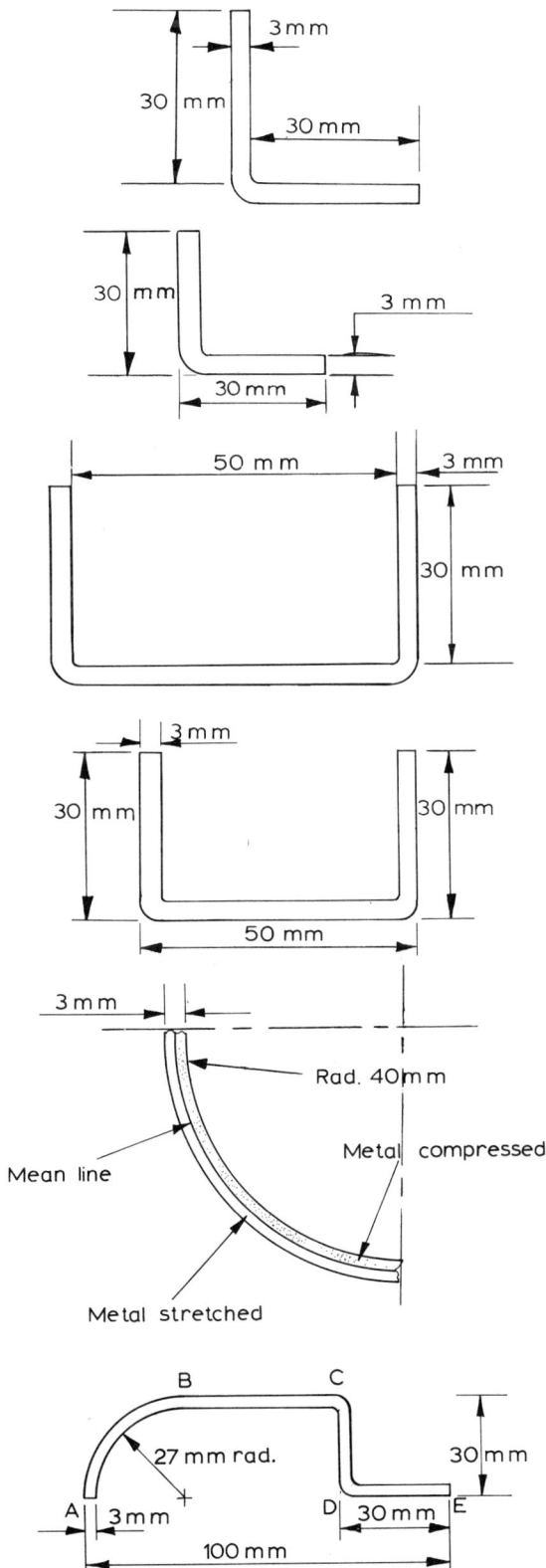

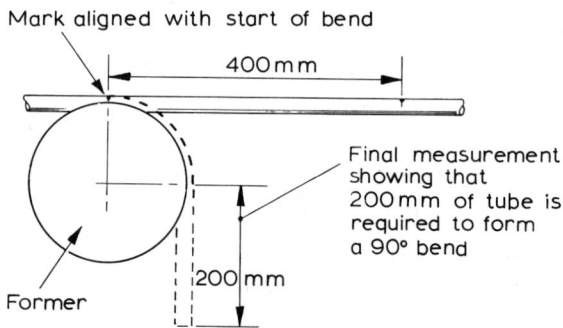

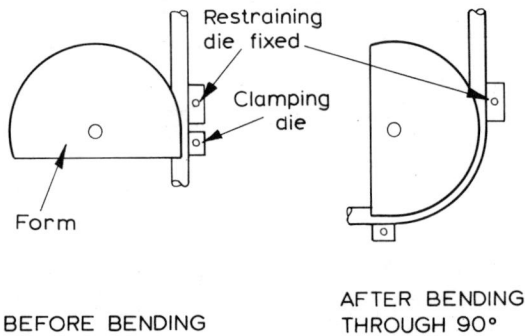

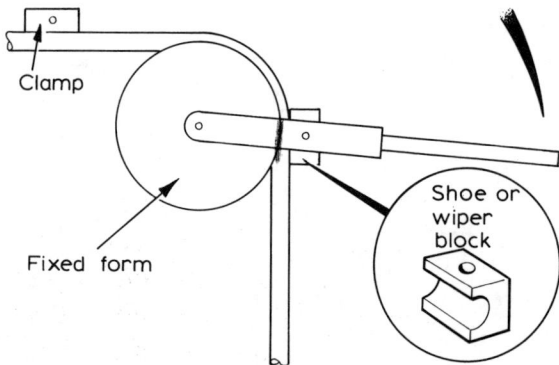

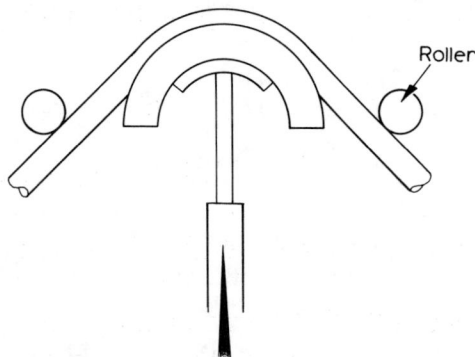

Pipe and Conduit Bending

When bending pipes and conduit there is a danger of the pipe wall collapsing as bending takes place. To eliminate this, bending should take place around a suitable former in a bending machine.

As the pipe will take up the radius of the former, it is not necessary to perform lengthy calculations to obtain the required cutting length, but to mark off a convenient length of pipe, perform the bend from the first mark and measure to find the amount of material used to form the bend (see top illustration).

The principle of draw bending is shown in the second illustration.

Mark aligned with start of bend

400 mm

Final measurement showing that 200 mm of tube is required to form a 90° bend

200 mm

Former

Restraining die fixed

Clamping die

Form

BEFORE BENDING

AFTER BENDING THROUGH 90°

The principle of wrap or radial bending is shown opposite.

Clamp

Shoe or wiper block

Fixed form

The bottom illustration shows a typical hydraulic ram pipe bender (the framework holding the rollers to the ram has been removed for clarity).

Roller

Large or exceptionally thin wall tubes have to be filled to prevent the walls collapsing. The filler material may be: (a) bobbin and followers; (b) springs; (c) rosin; (d) sand; (e) low melting point alloy.

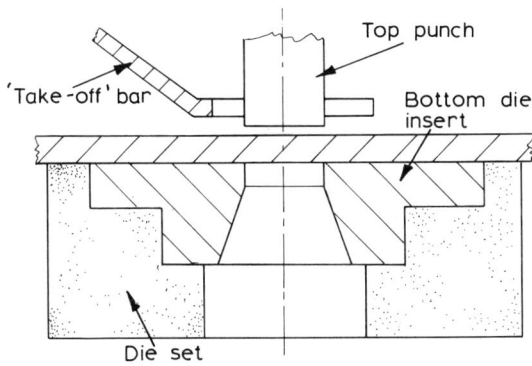

Die set

Take-off' bar — Top punch — Bottom die insert

Principles Involved in Punching Materials

Punching a hole in material requires a punch and die. The punch is positioned above the material which rests on the bottom die.

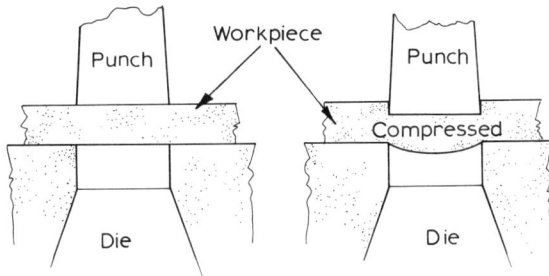

Punch — Workpiece — Punch — Compressed — Die

The punch, which is usually ground smooth to the correct diameter, descends under power and grips the material. The power stroke continues and the punch is forced into the material, compressing it. The depth of this penetration depends upon the hardness of the material and its thickness.

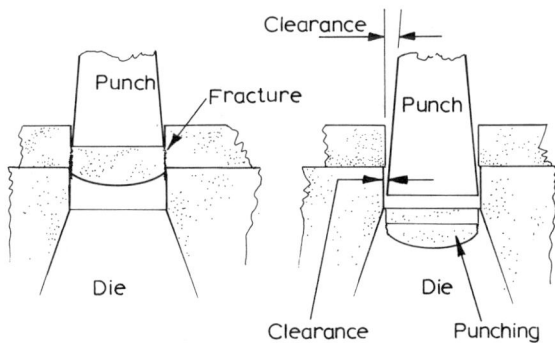

Punch — Fracture — Clearance — Punch — Die — Clearance — Punching

The material is now subjected to a very high concentration of stress in the area of the punch. This stress continues to build up until it is greater than the shearing or tearing strength of the material. When this occurs the material fails and the punch still under power descends through its full stroke and forces the ruptured metal or punching out of the hole, leaving behind it a torn hole which is larger on the underside of the plate than on the top.

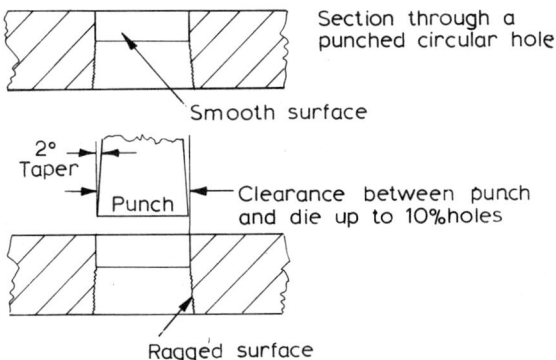

Section through a punched circular hole — Smooth surface — 2° Taper — Punch — Clearance between punch and die up to 10% holes — Ragged surface

The punch is now returned to the beginning of its stroke and, as it returns, has to be stripped from the material by a 'take-off' bar.

To reduce the power required to punch a hole, it is normal practice to have a 2° taper on the punch, and the bottom die-hole is made up 10% larger than the punch size. The machine may punch a single hole or a chain of holes; the latter type is called a 'piano punch'.

Principles Involved in the Forming of Metal by the Use of a Press Brake

The first important point regarding the bending of a strip of sheet metal is the consideration of the direction of the grain in the metal. Owing to the rolling process, by which sheet metal is produced, there is a grain in the metal running lengthways in the strip or sheet, not unlike the grain in a piece of timber.

A bend should, whenever possible, be across this grain rather than along it, as there will be less tendency for the metal to crack.

As most metals have a certain amount of natural spring the bend will have a tendency to open up after being bent, and to counteract this tendency a bending allowance is made on the angle of the top blade and the bottom vee die. For example, an 85° angle will be suitable for producing an angle of 90° on the workpiece.

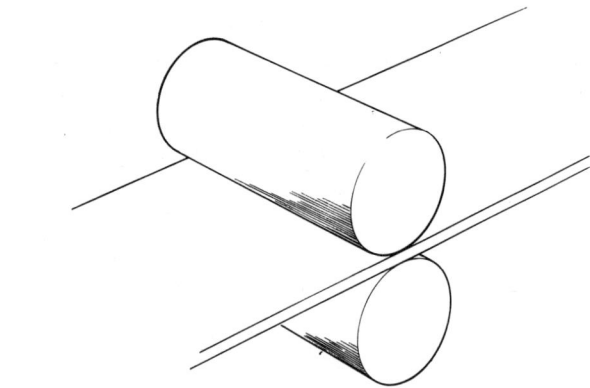

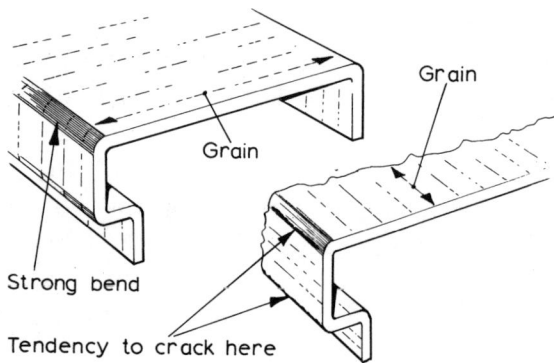

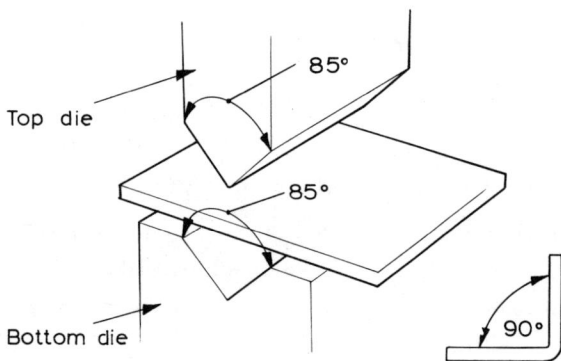

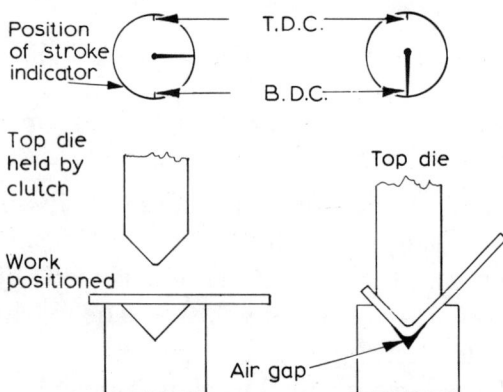

If the angle required is to be accurate, then by using an air-gap die and by altering the depth of the stroke on the press brake this can be achieved. Test-pieces are first checked for the correct angle before proceeding with the production quota. A press brake may be stopped at any part of its stroke by engaging the brake or clutch. This enables the operator to position the work under the blade.

E12   Templates   E12

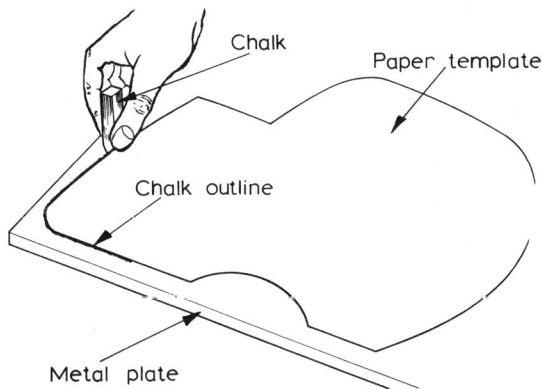

Chalk
Paper template
Chalk outline
Metal plate

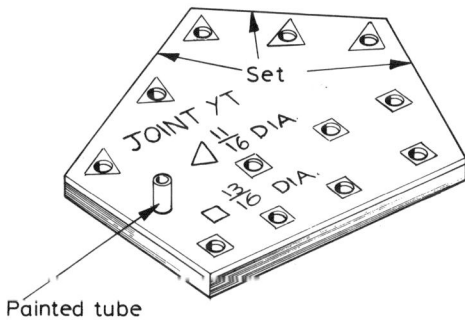

Set
JOINT YT
$\frac{11}{16}$ DIA
$\frac{13}{16}$ DIA.
Painted tube

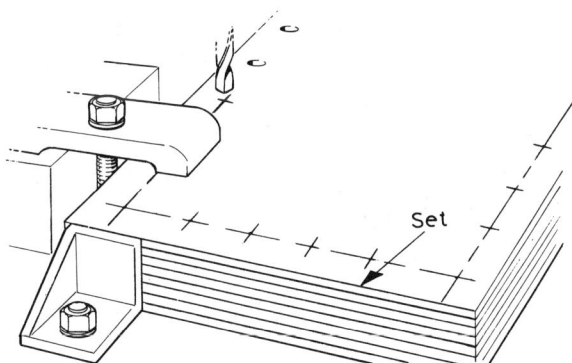

Set

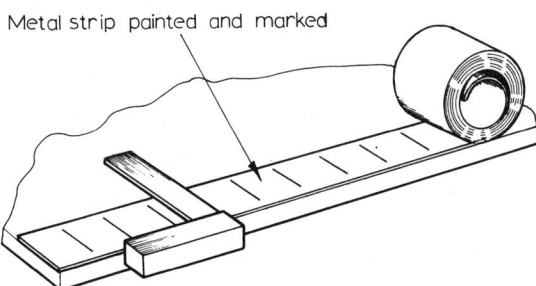

Metal strip painted and marked

Uses of the Various Types of Template

In structural work templates are made from paper, wood, hardboard, plastic and steel. If only one or two finished components are required then brown paper templates are used, as in illustration opposite, for chalking round before oxy-acetylene gas cutting to a required shape. If the job is to be repeated again then red template board is used for the template. This gives a longer life to the template.

Hardboard would be used as a drilling template. A small tube dipped in white lead (see opposite) is inserted into the holes in the template, leaving a ring of paint. The 'marker-off' then 'centre-pops' the centre of the ring before it is drilled.

If large steel plates are to be drilled then one of them is marked out accurately, positioned on the top of four or five similar plates, and clamped down. Using a radial drill the pile of plates is then drilled (pile drilling).

When finished plates are very large then strip templates of wood 100 mm × 15 mm thick are fastened together to the required shape (see opposite). Or the information is marked off on the finished plate from a painted strip of steel 50 mm × 2 mm thick which has been marked off previously in the template or loft shop.

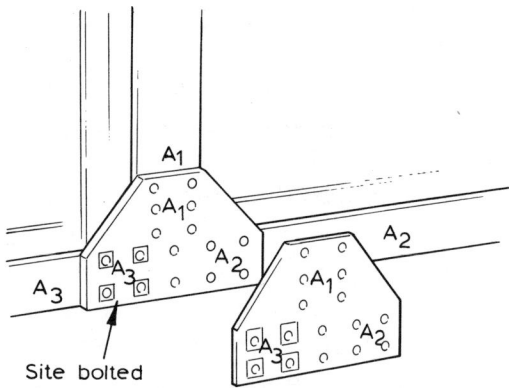

Site bolted

The Use of Templates for Fabricated Assemblies

Templates carry information on them for assembly procedure. An example of this is the square or triangular marking on a template. This indicates holes which are to be drilled, but the bolts are not to be inserted in them until final bolting-up on site.

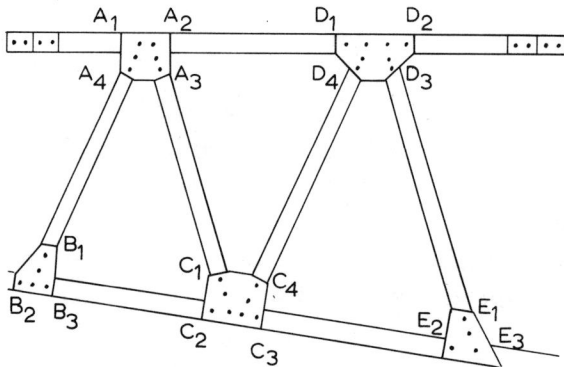

Templates may also indicate to the plater or structural worker that components are to be marked after processing—drilling, burning, etc., $X1$, $A1$, $G4$, for example—and these are to be assembled together. The site erection plan indicates where joints are positioned on the final assembly.

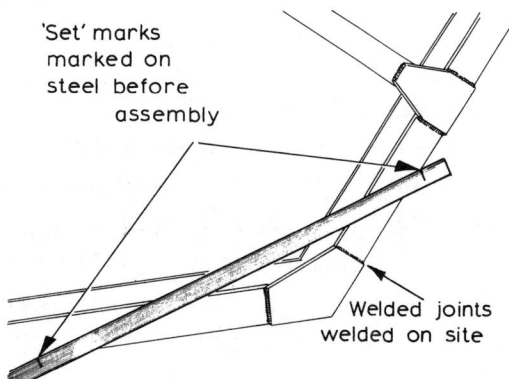

'Set' marks marked on steel before assembly

Welded joints welded on site

A metal strip template may be used on the final assembly to check bevels and the finished shape of three or more parts which are prefabricated on the site, before being lifted into position. This saves a lot of handling and readjustment when slung into its final position.

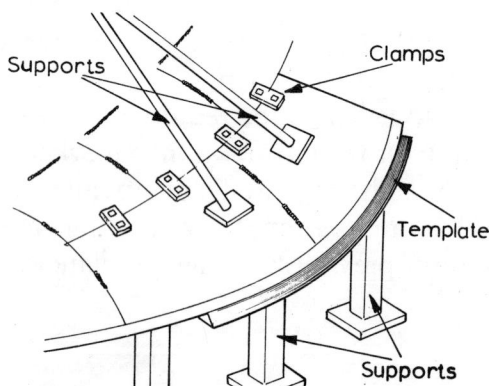

Supports

Clamps

Template

Supports

Templates are essential in the final assembly of hemispherical work. They may have to act as supports as well as holding and checking the final shape. Offset beams, cambered and bevelled work, all require accurate setting during the final assembly procedure.

E13   Testing and Colour Coding of Containers   E13

Testing of Containers by Air or Water

Containers, vessels, pipes, etc., are required to be tested to ensure that leakage will not occur during service. The pressure at which testing takes place would depend upon the conditions of service under which the article is to be used. For example:

Low pressure—a fuel tank containing a liquid not under pressure except that due to the weight of liquid in the tank.

High pressure—a pipe or pressure vessel, containing a liquid or gas, which requires a pressure to move the contents, as in a pipe, or to store or maintain a pressure, as in a steam boiler.

Three methods normally used for testing containers are filling with water, testing with air under pressure, and a hydraulic method. These are described below and over the page.

1   Filling with Water

This method would be used for open-topped containers such as dye vats, etc. It is important to remember that large quantities of water are extremely heavy and the test should be carried out either in the open or on a ground floor.

2   Testing with Air Pressure

This can be extremely dangerous—air can be highly compressed and the sudden release of this, due to failure of the vessel, can result in damage to the vessel if not serious injury to personnel. Vessels which require testing with compressed air at high pressure to simulate a live load should be sited in special buildings. Low pressure air testing, sufficient to simulate the weight of the material the tank would carry in service, is shown opposite. The tank is connected to a low pressure air-line. Any leaks are evident by the sight of bubbles in the water.

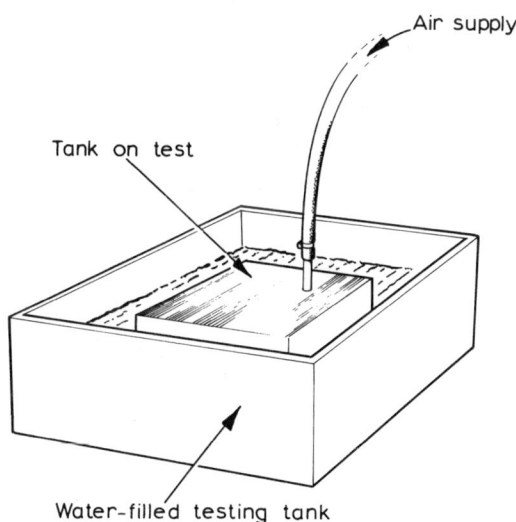

Air supply

Tank on test

Water-filled testing tank

3 Hydraulic Testing

All vessels requiring testing at medium or high pressure should be tested hydraulically. Any failure due to the pressure will result only in the loss of pressure and a slight spillage of water, as the water itself is not compressed but is merely a means of transmitting a force (see top illustration). If the force acting down on the small piston produces a pressure of 1 N/cm^2 in the pipe, this pressure acts upon each square centimetre of the large piston which, if this has an area of 100 cm^2, gives a total force of 100 N. By this means high pressures can be produced safely.

The illustration opposite shows a cylindrical vessel on test. The force acting down through the pipe can be:

(a) A head of pressure, i.e. the weight of the liquid in the pipe (suitable for low pressure testing).

(b) Connected to a pump (high pressure).

N.B. A vent pipe must be used to allow air to escape or it will be compressed giving rise to possible dangers, as mentioned on the previous page.

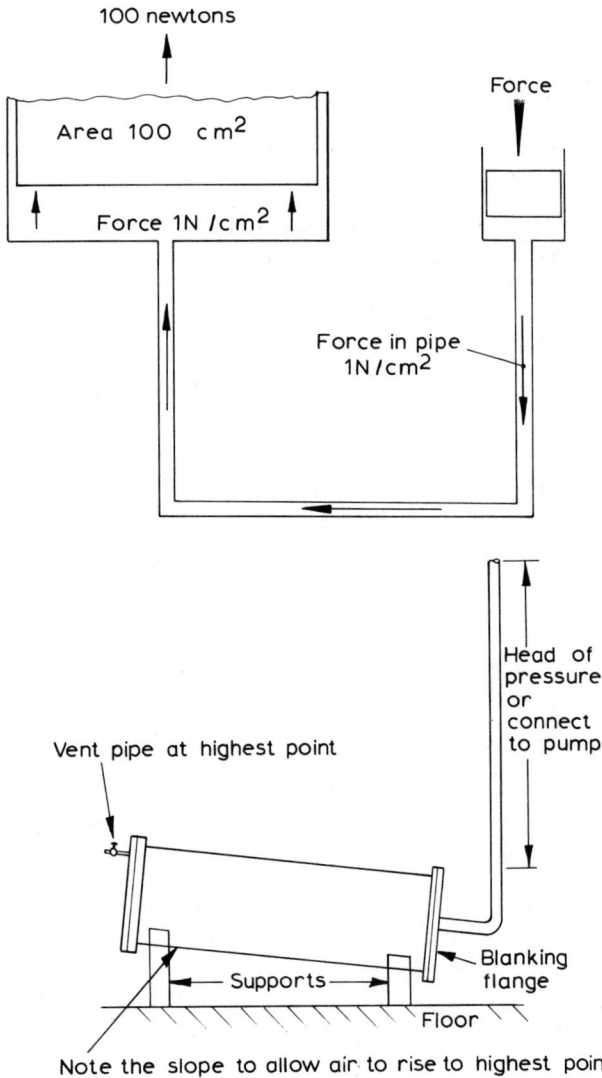

Colour Codes for the Contents of Pipes

British Standard 1710: 1960 lays down the colour coding used to denote the contents of pipes, although a revised Standard, based on I.S.O. R508, is under preparation. The use of a system of colour coding is most important as it enables the fitter to immediately locate a pipe carrying a particular service. The colour coding can appear on the pipes in one of the following five ways:

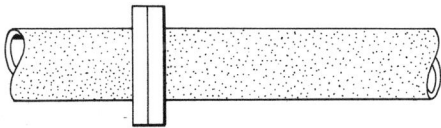

(a)   The whole pipe can be painted the correct colour.

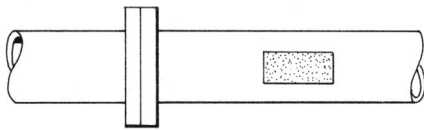

(b)   A panel can be painted on the pipe.

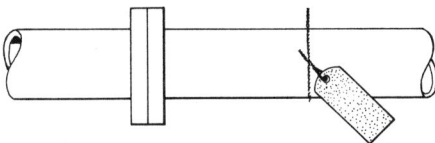

(c)   A coloured label can be attached to the pipe.

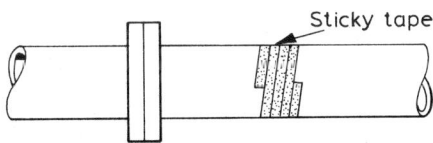

(d)   Coloured sticky tape can be fastened around the pipe.

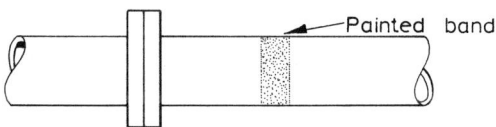

(e)   Coloured bands can be painted around the pipe.

The table below gives the colour codes used for denoting the different contents of pipes:

Colour	Contents
White	Compressed air
Black	Drainage
Dark grey	Refrigeration and Chemicals
Signal red	Fire
Crimson or Aluminium	Steam and Central heating
French blue	Water
Georgian green	Sea, River and Untreated water
Brilliant green	Cold-water services from storage tanks
Light orange	Electricity
Eau-de-nil	Domestic hot water
Light brown	Oil
Canary yellow	Gas

Section E   Fabrication and Welding

E   Test Questions   E

Section E1

1   What is a reinforced concrete beam?
2   Why is it dangerous to drill holes in prestressed concrete members?

Section E2

3   Sketch the following riveted joints:
    (a)   Single-riveted lap joint.
    (b)   Double-riveted lap joint.
    (c)   Double-riveted double cover-plate lap joint.
4   List the main causes of failure of riveted joints.
5   When are countersunk-head rivets used?
6   Where are (a) pan-head, (b) snap-head, and (c) mushroom-head rivets used?
7   What determines whether a black bolt or a fitted bolt is used?

Section E3

8   What is meant by (a) B.S.W., (b) B.S.F., (c) B.S.P., (d) B.A., (e) U.N.F., (f) U.N.C.?

Section E4

9   Sketch three ways by which an assembly made of thin sheet metal can be stiffened.
10   Why should joining beams be notched together?
11   When assembling a structure, if the holes of two joining members do not align correctly what procedure should be adopted?

Section E5

12   Sketch four different ways in which pipes can be joined.

Section E7

13   What is the colour of (a) an acetylene cylinder and (b) an oxygen cylinder?

14   Have acetylene attachments left- or right-hand threads?
15   What is the colour of (a) acetylene hoses and (b) oxygen hoses?
16   What is leftward welding and when is it used?
17   What is rightward welding and when is it used?
18   Explain the lighting-up and closing-down procedures for an oxy-acetylene welding set.
19   What is a gas regulator valve and how does it work?
20   What determines the power of a blowpipe?
21   Describe the difference between a neutral, a carburizing and an oxidizing flame. Give instances of where each is used.
22   Sketch six different types of welded joints.
23   What changes take place in mild steel during the oxy-acetylene welding process?

Section E8

24   Describe the difference between seam and spot welding.
25   What is meant by manual metal arc welding?
26   What are electrodes and with what are they coated?
27   What are the purposes of the electrode coating?
28   Describe the action of striking an arc.
29   What are the differences between a.c. and d.c. welding?
30   What is the danger of short circuiting?
31   How should electrodes be stored?
32   How are electrodes classified?
33   What happens to the weld when:
     (a)   the welding current is too high;
     (b)   the welding current is too low;
     (c)   the arc length is too great;
     (d)   the arc length is too small;
     (e)   the speed of travel of the electrode is too fast;
     (f)   the speed of travel of the electrode is too slow?

Section E   Fabrication and Welding

Section E9

34 What safety precautions should be observed when arc welding?

35 How should the eyes be protected when arc welding?

36 How are gas cylinders identified?

37 How should oxygen and acetylene cylinders be stored?

38 What precautions should be observed when moving gas cylinders?

39 How should gas leaks be tested for on an oxy-acetylene welding set?

40 What is meant by the following terms when related to welding?
  (a) Lack of penetration.
  (b) Lack of fusion.
  (c) Porosity.
  (d) Slag inclusion.
  (e) Undercut.
  (f) Overlays.
  (g) Cracking.
  (h) Blowholes.
  (i) Burn through.
  (j) Excessive penetration.

41 What workshop tests can be applied to welds?

42 What is a macroetch examination?

43 How can expansion and contraction in a welded joint be catered for?

Section E10

44 Describe the oxy-gas cutting process.

45 What is gouging?

46 What is the major difference between the blowpipe and nozzles used for welding and cutting?

47 Sketch the appearance of the cut edge that is obtained under the following conditions:
  (a) Cutting speed too fast.
  (b) Cutting speed too slow.
  (c) Nozzle held too high.
  (d) Nozzle held too low.
  (e) Cutting oxygen pressure too high.
  (f) Too large a preheating flame.

Section E11

48 What causes the workpiece to oxidize and burn when forging?

49 Why should the scale be removed from the workpiece after removal from the firebed before forging commences?

50 What are dollies and drifts and how are they used?

51 What are spinning and drawing? Sketch a simple arrangement illustrating each process.

52 What are bend allowances and why are they necessary?

53 What precautions should be taken when bending pipes and conduit to prevent the walls collapsing?

54 Describe the cutting action of a punch.

55 Why should a bend be made across the grain of the metal rather than along it whenever possible?

Section E12

56 What are templates?

57 From what materials can templates be made? Give reasons for each of the different materials used.

58 What information is placed upon templates?

Section E13

59 Why is it dangerous to use compressed air to test containers? What precautions should be observed if this is necessary?

60 Why is it preferable to hydraulically test containers whenever possible?

61 Why is it necessary to colour-code pipes?

62 What are the methods by which the colour coding can be placed on pipes?

F5   Magnetism   F5

Magnetism

A magnet is a piece of metal which can attract to itself, and hold, pieces of iron. The invisible force that enables magnets to attract other objects is called magnetism.

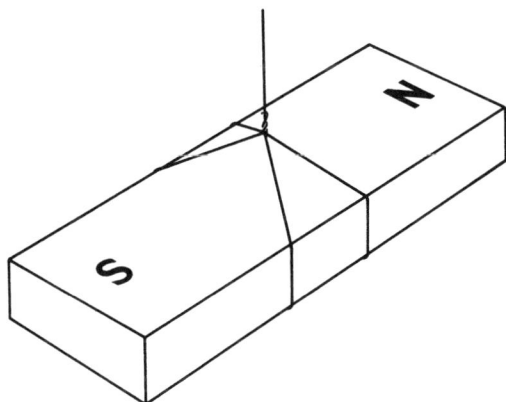

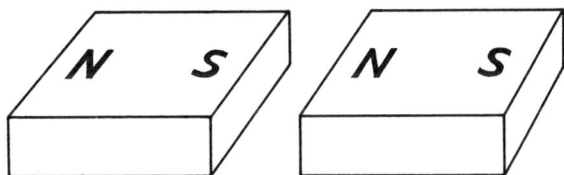

Poles of a Magnet

Any magnet which is suspended so that it is free to move, and is not affected by the field of another magnet, will always take up the same position in relation to the earth. One end of the magnet will always point North and is called the North-seeking or north pole of the magnet. The other end of the magnet is the South-seeking or south pole.

If a single bar magnet is cut in half the two pieces become separate complete magnets each having a north pole and a south pole.

ONE MAGNET MAKES TWO

Magnetic Lines of Force (Flux)

Magnetism cannot be seen but scientists picture it as a series of lines called magnetic lines of force or flux.

To see these lines carry out the following experiment using iron filings, a sheet of stiff paper and a bar magnet.

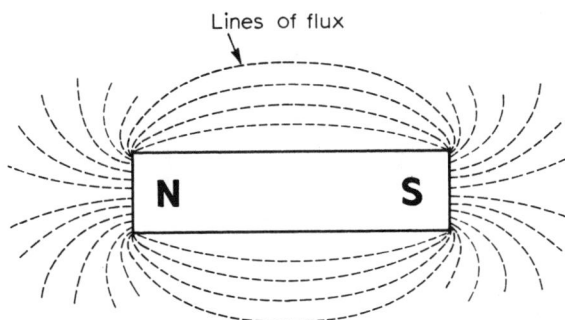

1 Place the paper over the magnet.

2 Sprinkle iron filings evenly over the paper.

3 Tap the paper with a pencil a number of times. The iron particles will now arrange themselves in the pattern shown here.

Magnetic Fields

The area near a magnet in which lines of magnetic force act is called a magnetic field.

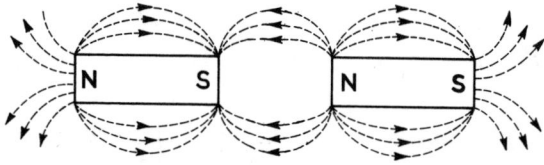

When the north pole of one magnet is near the south pole of another magnet then both magnets are attracted to each other.

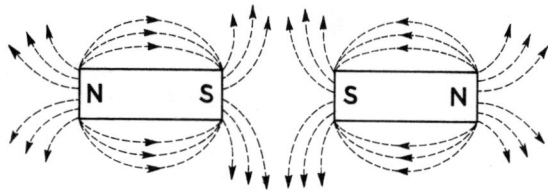

When two south poles or two north poles are nearest to each other, the lines of force turn away from each other.

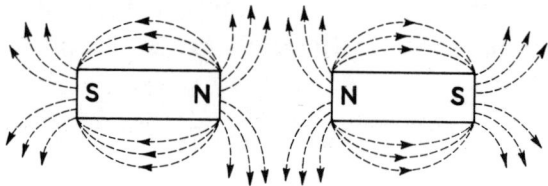

The magnets are pushed away from each other.

Law of Magnetic Attraction

From the above we can see that like poles repel each other and unlike poles attract each other.

Magnetic Materials

A material is said to be magnetic if it is attracted to a magnet.

The main magnetic materials are the metals, iron, nickel and cobalt.

Alnico—an alloy containing aluminium, nickel, iron, cobalt and copper—is used to make permanent magnets.

F6   Sources of Supply   F6

In the basic book we learned that a source of supply was necessary to drive electric current round a circuit.

D.C. Sources

Primary Cell

A primary cell produces electrical energy by chemical changes which take place within it. The chemical action is non-reversible which means that once the chemical is spent the cell is useless (i.e. it cannot be recharged).

There are several types of primary cell available but the most common is the *Leclanché cell*, which is manufactured in either the *dry* or *wet* form.

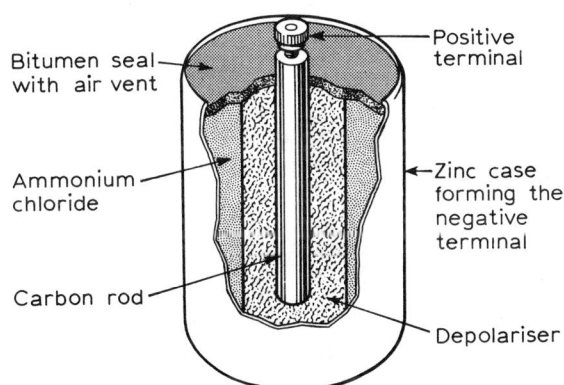

Dry Leclanché Cell

The dry cell, shown here, consists of a zinc case inside which is a rod of carbon packed round with a mixture of carbon powder and manganese dioxide. This mixture is then surrounded with ammonium chloride.

The function of the carbon powder and manganese dioxide is to act as the *depolarizer* and prevent hydrogen forming on the carbon rod.

The chemical reaction between the ammonium chloride and zinc case causes a current to flow.

This cell has an e.m.f. of approximately 1·5 volts when new.

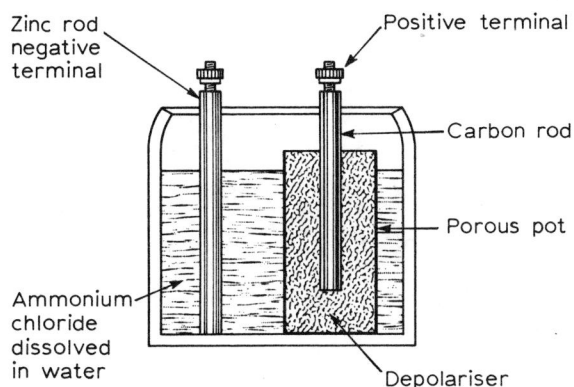

Wet Leclanché Cell

The wet cell has a carbon rod and depolarizer enclosed in a porous pot. A zinc rod and the porous pot are immersed in a solution of ammonium chloride and water.

This type of cell which is little used today has an e.m.f., when fully charged, of about 1·4 volts and when discharged this falls to about 1 volt. The specific gravity gives no indication of the state of this cell since it is about the same (1·17) whether charged or discharged.

The electrical circuit symbol for a cell is shown opposite.

Internal Resistance of a Cell

All cells offer resistance to the current flowing through them and this is called the *internal resistance* of the cell.

E.M.F. and the Effect of the Internal Resistance of the Cell

The voltage across a cell or battery when current is *not* flowing is called the electromotive force (e.m.f., see page 151 in the Base book).

When current is flowing the voltage available at the terminals of the cell will be less than the e.m.f. This is because part of the e.m.f. is needed to overcome the internal resistance of the cell.

$$\left(\begin{array}{c}\text{Voltage available at}\\\text{the terminals of a cell}\end{array}\right) = \text{e.m.f. of cell} - \left(\begin{array}{c}\text{voltage drop due to the}\\\text{internal resistance to the cell}\end{array}\right)$$

This is usually written as:

Terminal p.d. = e.m.f. of cell−(current flowing × internal resistance of cell).

The voltage drop is usually very small since cells are designed to have a small internal resistance.

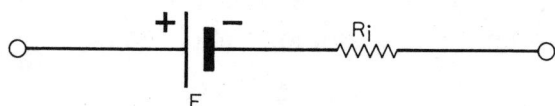

This diagram shows the circuit symbol for a cell of e.m.f. E and internal resistance R_i.

Cells connected in Series

This method is used to obtain higher voltages than a single cell can produce.

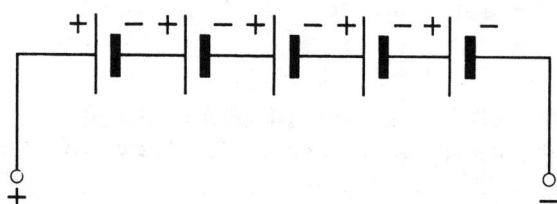

This is a circuit diagram of five cells connected in series, i.e. connected end to end.

The result is a battery of cells in series with one positive and one negative terminal.

The output voltage of the battery is the sum of the output voltages of each cell.

Example

A motor-car battery has six cells in series with each cell producing 2 volts. Find the output voltage of the battery.

Output voltage of battery = sum of the output voltage of each cell

= 2 V+2 V+2 V+2 V+2 V+2 V

= 12 V

Cells connected in Parallel

This method is used when a current value larger than a single cell can supply is required.

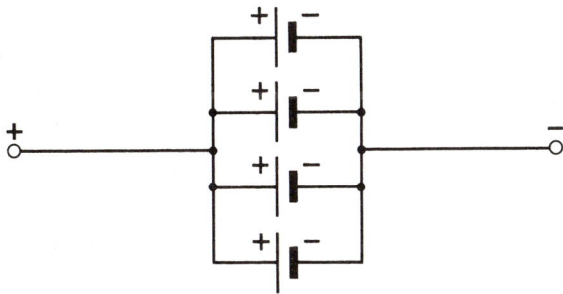

This is a circuit diagram of four cells connected in parallel.

The result is a battery of cells in parallel.

If the cells are similar, the output voltage of the battery will be the same as that of a single cell.

The current supplied by each cell will be a fraction of the current demanded by the load.

Secondary Cells (or Accumulators)

A secondary cell produces electrical energy by chemical action which is reversible. This means that the cell can be recharged by passing a current through it from another source.

There are two main types of secondary cell, *lead–acid* and *nickel–cadmium* cells.

Lead–Acid Cell (Accumulator)

This cell consists of two kinds of lead plates, one called the positive plate and the other the negative. It is made up of alternative positive and negative plates immersed in the electrolyte, which is a solution of sulphuric acid and distilled water.

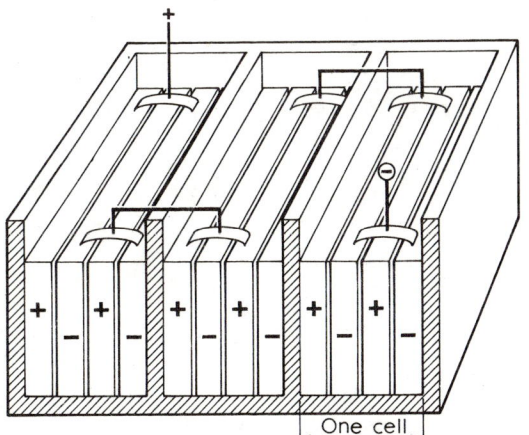

LEAD ACID BATTERY CONSISTING OF THREE CELLS IN SERIES

Negative plate—the holes in the plate are filled with spongy lead.

Positive plate—the holes in the plate are filled with lead peroxide.

Both the lead and lead peroxide react with sulphuric acid to form lead sulphate.

When the cell is run down both sets of plates are covered with a white deposit of lead sulphate.

The recharging current turns lead sulphate back to lead on one plate and lead peroxide on the other.

The e.m.f. of this type of cell, when fully charged, is just over 2 volts. When supplying current, this falls to about 2 volts and when discharged is about 1·8 volts.

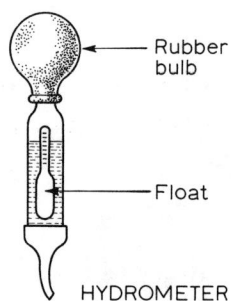

The state of the battery can be checked by measuring the specific gravity of the electrolyte with a hydrometer (left). When fully charged, the specific gravity is 1·25 (i.e. 1·25 times as dense as water) and when discharged, this falls to about 1·1.

Nickel–Cadmium Cell (Alkaline Cell)

This cell also consists of two sets of plates, one positive and one negative, but these are immersed in an electrolyte of potassium hydrate in distilled water.

Both sets of plates are constructed from steel having a series of holes in which are placed the active chemicals. The positive plates hold a mixture of nickel hydroxide and graphite, whilst the negative plates hold a mixture of cadmium and iron oxides.

F7   Consumer Appliances   F7

In the Base book (Electrical Power and Energy) we learned that:

$$\text{Power in watts} = \text{voltage} \times \text{current in amperes}$$

The resistance of the heating element of an appliance is also related to the power of the appliance by the formula:

$$\text{Power in watts} = \frac{(\text{Voltage})^2}{\text{Resistance of the element in Ohms}}$$

This may be rewritten as:

$$\text{Resistance of the element in ohms} = \frac{(\text{Voltage})^2}{\text{Power in Watts}}$$

Example

If an electric radiator has a rating of 1 kW and works off a 240 V supply, find the resistance of the element.

$$\text{Resistance in ohms} = \frac{\text{Voltage} \times \text{Voltage}}{\text{Power in Watts}}$$

$$= \frac{240 \times 240}{1\,000}$$

$$= \underline{57 \cdot 6 \text{ ohms.}}$$

Example

If a 240 V electric blanket has a rating of 100 watts, find the resistance of the element.

$$\text{Resistance in ohms} = \frac{\text{Voltage} \times \text{Voltage}}{\text{Power in Watts}}$$

$$= \frac{240 \times 240}{100}$$

$$= \underline{576 \text{ ohms.}}$$

For appliances working off the same voltage supply the smaller the power rating of the appliance the greater the resistance value of the element. This fact is shown by comparing the two examples above.

Current taken by an Appliance

For an appliance having a large power rating the current consumed is high because it has an element or resistance of low resistance value.

For an appliance having a small power rating the current consumed is low because it has an element or resistance of high resistance value.

Protection of Appliances by Fused Plug Heads

Appliances may be protected from damage due to excess current flowing by having a fuse in the plug which connects it to the supply (see diagram).

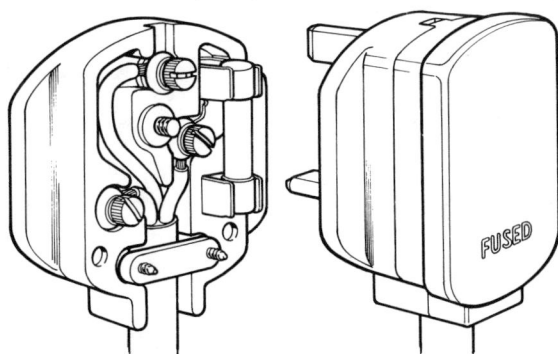

INTERIOR SHOWING FUSE   COMPLETE

Size of Fuse

The rating of the fuse used will depend on the power of the appliance.

Rating of Lamps

Lamps are rated by their power output, e.g. 60 W, 150 W, although this does not give a true indication of the light output of the lamp.

For instance, two lamps may have the same power consumption but different light outputs.

Lamps connected in Parallel

Lamps having the same voltage rating as the supply voltage are connected in *parallel*.

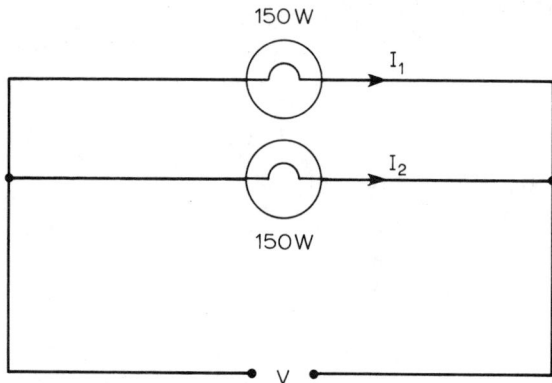

Connected in this way they will all take their normal operating current and so give their rated power output.

This illustration shows the circuit diagram for two lamps connected in parallel. The total current taken from the supply is the sum of the currents of the individual lamps.

The total power is the sum of the individual power ratings of the lamps.

Total current $= I_1 + I_2$

Total power in watts $= 150 + 150$

                    $= 300$ W.

Lamps connected in Series

Lamps of equal rating may be connected in series if the voltage of the lamps is lower than the supply voltage.

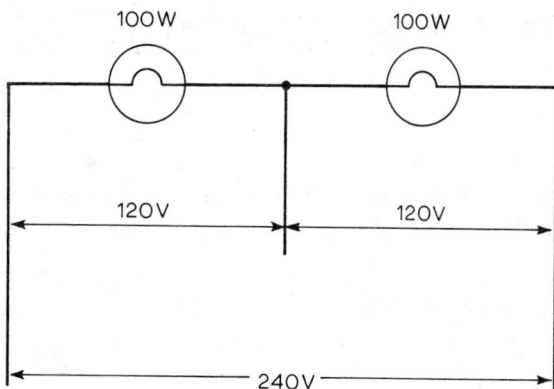

For example, two 100 watt, 120 volt lamps can be connected in series to a 240 volt supply. Both lamps will then give their rated power of 100 watts and take their normal operating current (see diagram).

A disadvantage of connecting lamps in series is that if any one lamp fails the others will not light.

F8   Terminations, Connections and Symbols   F8

Terminations and Connections

A termination is the entry of a cable end into an accessory of an electric circuit.

The method of preparing a cable for termination is as follows:

1   The excess cable is cut off with a suitable tool, e.g. pliers.

Outer sheath of insulation removed

Do not cut back
more than is necessary

2   The insulation is stripped off with a cutting knife no farther than is necessary, taking care not to damage the conductors.

Cut back to prevent
leakage from live parts

3   The outer protective sheath is cut back from the insulation by approximately 12 mm.

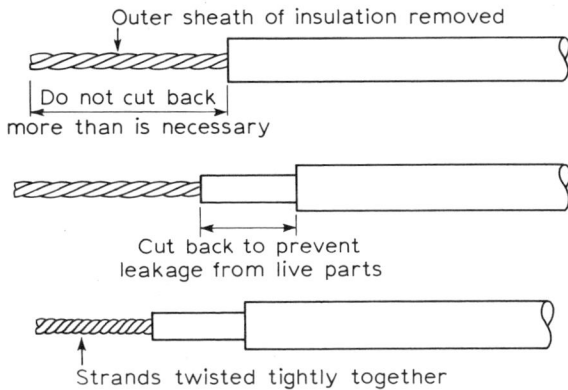

Strands twisted tightly together

If the cable has stranded conductors they should be twisted tightly together, care being taken not to damage the conductors.

It is required by the I.E.E. Regulations that a cable termination of any kind shall securely anchor *all* the wires of the conductor and not place any appreciable mechanical stress on the terminal.

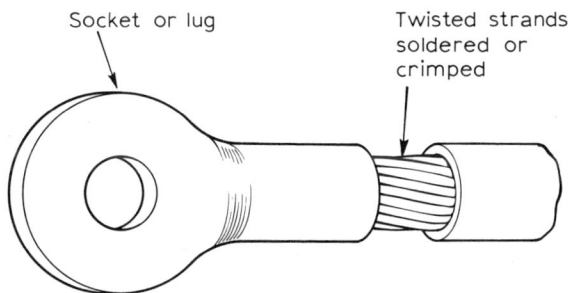

Socket or lug

Twisted strands soldered or crimped

If a soldering socket (or lug) as shown here is used, all the strands of the conductor must be contained. The reasons for this is to prevent the possibility of overheating and disconnection.

When using conduit, ducting or trunking, the cores of sheathed or unsheathed cable should be enclosed in a box or accessory of incombustible material.

Brass washer

Securing nut

Brass washers

Conductors

Conductor

Rubber insulated cables when terminated should not be exposed to direct sunlight.

There are a number of different types of terminal used in electric circuits. Some typical terminals are shown opposite.

Hole for conductors

TYPICAL TERMINATIONS

The brass pillar terminal is found in plugs, sockets and lampholders. This has a hole drilled through it in which the conductor is inserted and firmly held by a brass set screw. If the conductor is small compared with the hole, it should be bent over to make a better electrical contact.

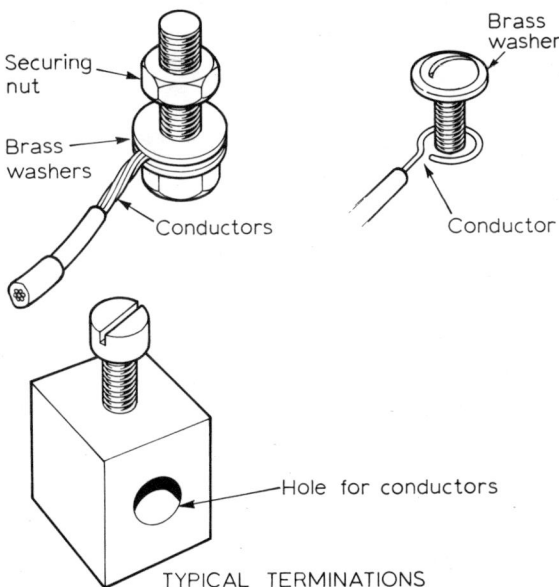

135

Colour Coding for Electrical Wiring

All cables whether single-core or multicore must be identifiable at their terminations and preferably throughout their length. This helps the electrician to select the correct conductor for connecting to a particular terminal on the appliance.

Rubber and p.v.c. insulated cable should have the insulation of each conductor coloured along its length in accordance with Table B4 shown in the I.E.E. *Regulations for the Electrical Equipment of Buildings*.

Sleeves or discs of the appropriate colour may be used at the terminations.

Three-core flexible cords have the following identifying colours:

Conductor	*Colour*
Live	Brown
Neutral	Blue
Earth	Green/Yellow

Four-core flexible cords have the following colours:

Conductor	*Colour*
Three Live	Brown
Earth	Blue

Conventional Symbols for Common Electrical Components

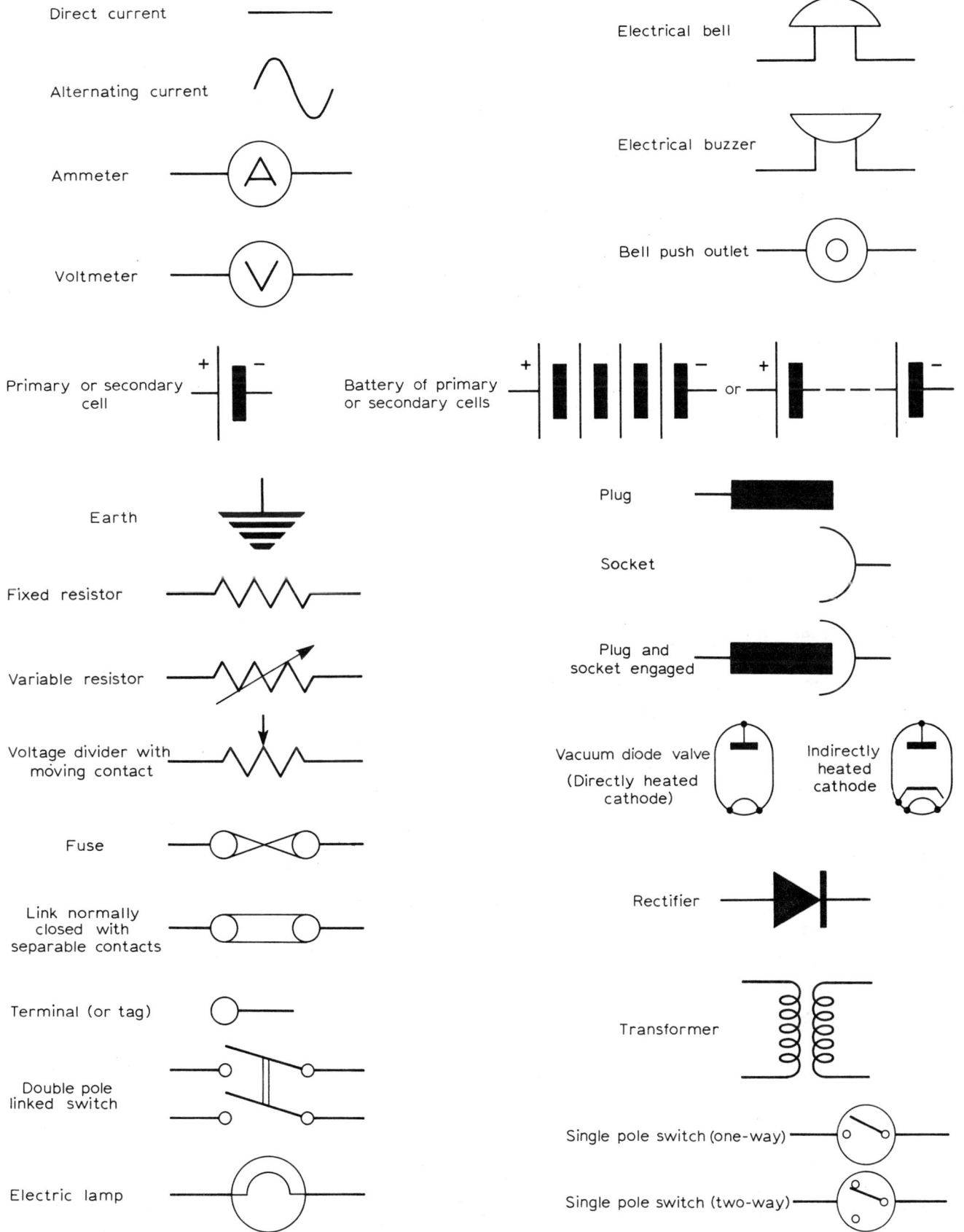

Direct current

Alternating current

Ammeter

Voltmeter

Primary or secondary cell

Earth

Fixed resistor

Variable resistor

Voltage divider with moving contact

Fuse

Link normally closed with separable contacts

Terminal (or tag)

Double pole linked switch

Electric lamp

Electrical bell

Electrical buzzer

Bell push outlet

Battery of primary or secondary cells   or

Plug

Socket

Plug and socket engaged

Vacuum diode valve (Directly heated cathode)

Indirectly heated cathode

Rectifier

Transformer

Single pole switch (one-way)

Single pole switch (two-way)

F9   Wiring Regulations F9

Necessity for Wiring Regulations

In order that there shall be no danger from electric shock or fire the Institution of Electrical Engineers have a publication called *Regulations for the Electrical Equipment of Buildings.*

These regulations form the basis of all good practice for electricians with particular reference to the electrical contracting industry.

Protection against Shock and Fire

One of the main methods of protecting against shock and fire is to connect all non-current-carrying metal work to earth. This prevents a dangerous build-up of voltage on an appliance or metal work.

Excess Current Protection

To prevent damage to appliances and possibility of fire, all circuits must be disconnected from the supply automatically if too great a current flows.

This is done by using either a fuse or excess-current circuit breaker.

In both cases, if the protective device operates, the reason for its operation must be found before the circuit is re-connected to the supply.

Current Leakage Protection

A fuse will also protect a circuit from current leakage if the earthing is adequate. Where the earthing is not adequate, special earth leakage circuit breakers have to be used.

These are devices which operate when 40 volts are developed by the leakage current across the circuit-breaker coil.

Protection against Corrosion

Electrical plant and cables must be protected when used in corrosive situations either by using galvanized conduit or having cables with a p.v.c. sheathing over them. Appliances may have to be painted with an anti-acid paint.

Protection against Mechanical Damage

All electrical plant, appliances, cables and wires must be adequately protected from mechanical damage.

Earth Continuity Conductor

Where earthing is used as a means of protection, all non-current-carrying metal work must be connected back to the earth terminal at the supply intake point. This is called an earth continuity conductor and when it takes the form of a separate conductor it must be insulated and coloured green.

Fuses

There are two types of fuses:
1 Rewirable type;
2 Enclosed type or cartridge fuse.

1 Rewirable-type fuse

This is a piece of wire connected to contacts on a fuse link which fit into a fuse base. The wire, which is called a fuse element, may be contained in an asbestos tube.

2 Cartridge-type fuse

The fuse element is sealed into a ceramic tube with contacts formed on the ends of the tube.

Rewirable fuses have the following disadvantages:
(i)   Hot metal may be scattered when the fuse operates.
(ii)   It is possible to fit an incorrect size of wire.
(iii) The fuse element may age and therefore deteriorate.

Cartridge fuses do not have these disadvantages, but are more expensive.

Rating of Fuses

The current rating of the fuse is the current which the element will carry without deterioration.

Fusing Factor

A fuse element will operate at approximately twice the value of the fuse rating depending on the type and construction.

Miniature Circuit Breakers

These are devices which disconnect the circuit or appliance from the supply either by magnetic or thermal trips. They are small and are designed to fit into consumer units for domestic circuit protection.

Coarse Excess Current Protection

A device which will not operate within four hours at 1·5 times the load current of the circuit.

Close Excess Current Protection

A device which will operate within four hours at 1·5 times the load current of the circuit.

Supply Voltages

The standard domestic supply voltage in the U.K. is 240 V a.c. Industrial supplies for general distribution have a voltage of 415 V a.c. Higher voltages are available for special installations.

F   Test Questions   F

Section F5

1   Sketch the magnetic field that would result from a bar magnet being cut into two pieces.
2   Sketch the magnetic field of two south poles which are close together.
3   Describe how the four main points of the compass may be determined with the aid of a simple bar magnet.

Section F6

4   A cell has an electromotive force of 2·0 V and internal resistance of 0·2 ohm. Determine the voltage at the terminals when the cell supplies a current of 1 A.
5   State the reasons for connecting cells in:
    (a)   series;
    (b)   parallel.
    Make a sketch of each arrangement.
6   Six cells of 1·8 V each are connected in series to form a battery. Find the resultant voltage between the battery terminals.
7   What is the basic difference between primary and secondary cells with reference to the chemical action?
8   State the name of the chemical which forms the electrolyte of:
    (a)   a lead–acid cell;
    (b)   an alkaline cell.
9   Briefly describe with the aid of a sketch the construction of a dry cell.
10   State the name and make a sketch of the instrument used to measure the specific gravity of the electrolyte used in a secondary cell.
11   Give the factors which indicate the condition of charge of a secondary cell.

Section F7

12   Calculate the current taken by an electric heater having a rating of 3·5 kW when working on its rated voltage of 240 V.
13   Determine the resistance of a 750 W 250 V lamp when it is operating at its rated power output.

14   Determine the supply voltage of a 2·5 kW heater, having a resistance of 16 ohms.
15   Explain how the fuse in a plug-head will protect an appliance from damage.
16   Why is it dangerous to replace a fuse from a plug-head by a wire fuse?
17   Describe briefly two devices which are used to protect electrical appliances from damage due to excess currents.
18   What are the advantages of cartridge fuses over rewirable fuses?
19   A fuse has a rating of 30 A. Explain what this means.

Section F8

20   Make neat diagrams which show the conventional symbols of the following electrical components:
    (a)   Earth;
    (b)   Variable resistor;
    (c)   Fuse;
    (d)   Cell.
21   Make a circuit diagram, using conventional symbols, of a battery supplying current to an electric lamp. Include in the diagram a fuse, an ammeter and a voltmeter.
22   What is a termination?
23   Describe how you would prepare a cable for termination.
24   Why must all the strands of a conductor be contained in a terminal lug?
25   What special precaution must be observed when terminating rubber insulated cables?

Section F9

26   How are conduits protected from damage due to corrosion?
27   Explain how the earth continuity conductor of an installation will protect persons from the dangers of electric shock.
28   Explain the terms close and coarse protection when applied to electrical installations.

29 Explain why an armoured cable must be protected from mechanical damage. Suggest how this could be done.

30 Why is it necessary to protect an installation or appliance from earth leakage currents?

31 Describe two devices that will protect an appliance from the dangers of earth leakage currents.